Migration and Education in a Multicultural World

Migration and Education in a Multicultural World
Culture, Loss, and Identity

Ursula A. Kelly

MIGRATION AND EDUCATION IN A MULTICULTURAL WORLD
Copyright © Ursula A. Kelly, 2009.

All rights reserved.

First published in 2009 by PALGRAVE MACMILLAN® in the United States—a division of St. Martin's Press LLC, 175 Fifth Avenue, New York, NY 10010.

Where this book is distributed in the UK, Europe and the rest of the world, this is by Palgrave Macmillan, a division of Macmillan Publishers Limited, registered in England, company number 785998, of Houndmills, Basingstoke, Hampshire RG21 6XS.

Palgrave Macmillan is the global academic imprint of the above companies and has companies and representatives throughout the world.

Palgrave® and Macmillan® are registered trademarks in the United States, the United Kingdom, Europe and other countries.

ISBN-13: 978-0-230-61292-1
ISBN-10: 0-230-61292-X

Library of Congress Cataloging-in-Publication Data

Kelly, Ursula Anne Margaret, 1956–
  Migration and education in a multicultural world : culture, loss, and identity / Ursula A. Kelly.
    p. cm.
  Includes bibliographical references and index.
  ISBN 0-230-61292-X
  1. Social change. 2. Emigration and immigration. 3. Education. I. Title.

HM831.K46 2009
306.4'209718—dc22   2008033837

Design by Scribe Inc.

First Edition: March 2009

A catalogue record of the book is available from the British Library.

First edition: March 2009

10 9 8 7 6 5 4 3 2 1

Printed in the United States of America.

The foundational essay "The Place of Reparation: Love, Loss, Ambivalence and Teaching," in *Teaching, Learning and Loving*, edited by Daniel Liston and Jim Garrison (RoutledgeFalmer, 2004), is considerably revised and extended as Chapter 5 of this book. While no permission is necessary, I would like to acknowledge RoutledgeFalmer.

The poem, "Bury me in Newfoundland," is reprinted with permission of the author, Danielle Devereaux.

Transferred to Digital Printing in 2010

To the memory of my parents
Margaret Waterman Kelly and Andrew J. Kelly

Contents

Acknowledgments	ix
Introduction: At Sea: Toward an Educational Discourse of Loss and Place	1
1 Losing Place: Reluctant Leavings and Ambivalent Returns	23
2 Writing "The Distance Home": Migration, Mourning, and Difference	47
3 "The Word, for Loss": Literacy, Longing and Belonging	71
4 Separation, (Re)connection, and a Transformative Education of Place	93
5 The Place of Reparation: Loss, Ambivalence and Teaching	121
Conclusion: "Learning to Live with Ghosts": Loss, Place, and Education	147
Notes	169
References	175
Index	185

Acknowledgments

The impetus for this book comes from a love of a place and its people that attempts to resist romanticism and the allure of tradition in order to see cultural criticism as an opportunity for contemplation and growth. Its inspiration are parents who believed that high expectations and responsible action were necessary companions to love of any sort. It is to them I am most, and most often, thankful.

My graduate students in the Faculty of Education at Memorial University of Newfoundland, most of whom have spent many more years living and working in this province than I, have taught me much about love, commitment, loss, and resilience. I have been deeply moved by the ways in which they have opened their hearts to who I am and to the work we do together. I feel embraced and welcomed by them—at home. They are among the greatest gifts of this place and, as educators, the source of its greatest hope.

Alice Collins, the Dean of the Faculty of Education during the time in which this book was conceptualized and written, supported my writing by, in numerous ways, supporting me. I am most thankful for her gracious gestures.

While leaving was the experiential rupture at the heart of this book, it was returning that brought creative insight and renewal. Clar Doyle, a fine scholar, teacher, colleague, artist, and friend, was instrumental in my decision to come home. In my heart, I thank him daily for the support that helped make my return possible.

Pat Singer, first reader and life partner, lives with the expressed vicissitudes of my love of this place and, through a shared love and an unfailing support, ensures that I, and we, endure and thrive.

Bury me in Newfoundland

Don't bother with a box. Lay me
right in the rock, naked. Put a marker

if you must. On a piece of driftwood
burn my name but know

I won't be here long. Last fistful of dirt
and I'm gone. Bone splinters shredding skin.

Shards of me marble the dark gray granite. I am
jagged cliff edge. Spools of hair tumble, reach down

along the sharp red sand, under the smooth backs
of beach rocks. I am seaweed, luscious green

licking salt from the sea. My teeth shatter
on the rock below, washed in white foam,

rubbed with salt. I am broken seashells,
scraping and pulling in time with the tide.

My breasts find a barren, caress
the carpet of moss till nipples

push through—partridgeberry,
the rich, red sting.

—Danielle Devereaux

Introduction

At Sea[1]
Toward an Educational Discourse of Loss and Place

An Introduction

The meanings of place, home and belonging are highly contested in contemporary times marked by heightened movement, dislocation, contested borders, and multiple diasporas. Arising from the devastating changes that are part of the legacy of the twentieth century—unprecedented worldwide migration, unrelenting global conflict and warring, unchecked materialist consumption, and unconscionable environmental degradation—are questions about the toll of loss such changes exact. As large scale and ubiquitous as these changes are, their specificity is also inarguable. Often set against a backdrop of transmigration, struggle and loss continue to be lived and experienced through deep specificity, in geospatial arrangements and with structured affect that continues to assert the importance of place as a critical construct. Attending to such specificity emphasizes the interconnections between contexts and broader movements and remains a prudent route to articulating these critical interconnections among places and people in complex times. In this spirit, then, this book turns to such specificity as a means to examine the inflections of change, loss, and identity—disorientation, melancholy, and difference as sites of regression and possibility—in order to discern the nature of a critical education framed by a concern for these conditions.

The specific context and cultural ethos to which such specificity is directed, in this book, is the Canadian province of Newfoundland and Labrador.[2] A largely White settler (former) colony,[3] its recent history has been marked by severe social and cultural challenges due to the widespread impact of globalization and ecological crisis. The turmoil and change that have resulted raise important questions related to the cultural politics of loss, belonging, and place. These particular circumstances, part of the emerging complexities of "new times" (Hall, 1989), which demand modes of discussion and analysis in ways that attend to the mobilization and effects of power as well as cultural ethos and affect, are not confined to this place. But an examination of the particularity and complexity of loss in this place has something to suggest about other places, too. Looking at loss as a threshold of possibility, this examination of place focuses on themes of migration,[4] melancholy, reparation, and education against a backdrop of what Kobena Mercer calls "multicultural normalization" (Mercer, 2000, p. 234) and the ethical demands it poses for places and for practices of conviviality (Gilroy, 2005). These central concerns to which each of the chapters will turn are outlined here in an introductory manner.

Culture, Loss, and Place

[T]here will always be that ache of loss about Newfoundland for me, a sense of how part of what it was is passing out of our lives for good.

Michael Crummey[5]

It may be that the dilemma of which I am speaking is so deep, freighted with such an absolutely foundational question—has the idea of Newfoundland as Newfoundlanders have always known it run its course?—and raises so large an anxiety that we are not ready to encounter it. Or there may be an even more gloomy possibility. That the decline is seen as inexorable, is too deep for repair, that what cannot be halted is best not spoken of.

Rex Murphy[6]

Taken from its aboriginal inhabitants and claimed and settled as a fishing resource-based colony of the British Empire over five hundred years ago, the once-independent nation and now Canadian Province of Newfoundland and Labrador, today, faces profound social, cultural, and ecological challenges in the face of the devastation and collapse of the cod fishery and, simultaneously, the development and production of an offshore oil and gas industry. The demise of the cod fishery has as its backdrop a crisis of the world's oceans of unprecedented proportions related to international overfishing, multiple species depletion, and the effects of climate change. The rise of the oil and gas industry has as its backdrop a world hunger for energy fueled by industrial overproduction and rampant consumerism, and accompanied by large scale indifference to the impact of postindustrial lifestyles on planetary resources. That both these resource crises have huge implications for citizens of the province is unarguable. Regardless of context, any local responses to change rarely keep pace with the immense impact of such large-scale crises. As the communities of Newfoundland and Labrador empty of the families of its fisheries workers in the hundreds of thousands and as entire communities disappear, anger, frustration, and despair—and the attendant symptoms of social crisis—have heightened.

In 1993, my first book, which addressed cultural politics in relation to Newfoundland and Labrador, was published, just one year following the Canadian government moratorium on the five-hundred-year-old cod fishery in the province. In that book, *Marketing Place*, I examined the intersections of region, culture, power, and identity to discern the possibilities of an educational project of cultural renewal and change premised on a notion of critical literacy, a "reading the world" in order to name inequity and to challenge and to usurp its prevailing assumptions and practices. Based on research conducted in the mid-1980s, *Marketing Place* holds no reference to the cultural catastrophe unfolding during the time of the research on which it is based. Nor does it in any way portend the watershed moments to follow—the initial moratorium of 1992 and the subsequent closure of the fishery in 2003. This gaping silence could be attributed, in part, to my by then decade-long absence from the Province. Undoubtedly, too, it could be related to a wider cultural attitude that took the sea and its bounty for granted and its place in our lives as incontrovertible. More profoundly, the silence could be attributed to economic presentism and the kinds of denials and sublimations that maintain the status quo in any context. This present book attempts to take stock, culturally and educationally, of what has come to pass in these ensuing fifteen years, its historical roots of affect, and present and future implications and possibilities.

The title of this introductory chapter, *At Sea*, refers to a shortened version of "all at sea," a phrase in use in the 1700s, prior to the development of advanced navigational instruments, to describe the precarious and uncertain state of a boat out of sight of land and in danger of becoming lost. The meaning, too, evolved so that the shortened version metaphorically describes a general loss of direction, of becoming unmoored and disoriented, without a secure position. This metaphor aptly describes, in a language in keeping with its history, geography and preoccupation, the transitional character of "these times" in Newfoundland and Labrador. Being "at sea" is not a position of inherent despair or lack of resilience, although, as in any transition, this position is one possibility. Similar to Nancy Huston's *Losing North*, a metaphor that flags loss, betrayal and disorientation (Huston, 2002, pp. 2–5), "at sea" captures profound disequilibrium at the level of individual identity and at the level of the cultural collective, as well.

Structures of Loss

A state of disequilibrium marks loss as a destabilizing, reorienting force. Indirection, like loss, offers an opportunity to reassess and envision differently, to ask pointedly of the implications of cultural loss and to inquire of pending change its direction and focus. It is these questions and their many dimensions that form the basis of the essays that comprise this book. *Migration and Education in a Multicultural World: Culture, Loss, and Identity* is a collection of essays that focuses on the personal, social, and psychic dimensions of cultural change. The essays are framed within cultural theory, aspects of psychoanalysis, and critical educational theory, and they are contextualized through an analysis of various texts of autobiography, memoir, and cultural narrative. The essays address issues of cultural crisis, loss, and hope—specifically, but not exclusively, as these issues relate to Newfoundland and Labrador—and they conceptualize broadly how education might be envisioned to respond productively to the challenges of loss and change.

Human migration, what John Berger (1972) called the quintessential experience of the past century, is often both a consequence and a doubling of loss. While gain can also accrue through migration, its severing moment from established home places, cultural practices, and identities throws into chaos any held beliefs about cultural identity as fixed or stable. With migration, identity is renegotiated through a visceral process of becoming in relation to new contexts, new challenges, and new impositions of versions of oneself and, and by, others. Roger Bromley (2000) refers to such identities as those dislodged through the experiences of migration as "identities at risk," which "represent a positive, if unsettling, phenomenon by opening up the possibilities of new affiliations" (p. 3). The extent to which such formations are hopeful, however, is contingent on the manner in which loss is confronted and processed.

Loss, itself, is multifaceted and, as a structure of feeling, it is extremely complex. Loss has many faces, individually and culturally. Culturally, nostalgia is one of our most common responses to loss. Overton (1996) reminds us that a long history of economic migration has fueled nostalgic relationships to the province of Newfoundland and Labrador by both expatriates and impatriates. This affliction of the exiled is as much a symptom of separation from a beloved home and community as it is one of displacement and alienation at the site of relocation. This particular, often romanticized, register of loss propels the desire of return, a sentiment of use in the context of uneven economic development and the fluctuations of labor supply and demand, nationally and globally (Overton, 1996; Kelly, 1993). The desire to return, where return is possible and feasible, as within nation states and among nations where political circumstances are not dire, allows for an often seasonal flow of labor in keeping with the demands and fluctuations of a global economy.

More profoundly, melancholia is a register of loss that suggests not just loss but the denial of this loss through its reinstatement in the ego. Here, *melancholia* is differentiated from the *melancholy* of everyday use, which denotes a general sadness. Melancholia is used here to describe a psychic mechanism of *identification* with what is loved and lost, the lost object, to the detriment or loss of one's self. As a result, as Cheng (2001) notes, melancholia is deeply ambivalent because, in melancholia, "the relationship to the object [the person, place, concept, or thing] is now no longer just love or nostalgia but also profound resentment" (p. 9). This complex, ambivalent dynamic creates cultural subjects of loss under conditions that are as much about what Cheng (2001, p. 20) calls "surviving grief as embodying it," that is, a negotiation of loss and an incorporation of what is lost, a symptom of the extent to which social hurt can shape collective identity in complex ways that are neither easily discernible nor obvious.

Finally, there is the register of loss as a result of ecological and environmental devastation and community displacement, such as that experienced by aboriginal peoples worldwide and, in Newfoundland and Labrador, that experienced by the Innu of Labrador as a result of hydro development and nickel mining. Albrecht (2005) has referred to the effects of such loss as *solastagia*. He argues that this inability to find solace in a beloved place "is manifest in an attack on one's sense of place, in the erosion of the sense of belonging (identity) to a particular place and a feeling of distress (psychological desolation) about its transformation" (Albrecht, 2005, p. 45). This distress has real consequences for individual and community health as solastagia is manifested as increased and diverse expressions of social dysfunction, despair, and aggression. Along with the province's aboriginal peoples, the symptoms of solstalgia mark rural areas of Newfoundland and Labrador, too, as it struggles with the dramatic decline of ocean life and historically unprecedented changes that are a consequence of an inability to sustain communities through a life of work on the ocean.

These various forms of grief require articulation as part of the process of confronting the vicissitudes of loss. Anne Anlin Cheng (2000, p. 29) connects clearly the importance of acknowledging loss and articulating grief to any viable project of social change. She writes, in relation to racial melancholia, specifically, that *"if we are willing to listen* [emphasis mine], the history of disarticulated grief is still speaking through the living, and the future of social transformation depends on how open we are to facing the intricacies and paradoxes of that grief and the passions that it bequeaths." Disarticulated grief has a particularity in relation to race, but disarticulated grief creates distortions, again, with deep particularity, on other sites of marginalization, such as gender, social class, and sexual identity, as well, as many scholars have argued (Butler, 1997; Eng & Kazanjian, 2003; Silin, 2006) and as the individual chapters of this book will demonstrate.

Newfoundland and Labrador has its own specific history of disarticulated grief. Yet, such specificity does not preclude a deep interconnection to and implication[7] in other and others' histories, the disarticulated grief of "others" as well as, and along with, its own. More than in a manner of speaking, we are always already the remains of numerous losses. But what is to be productively done with such remains of loss? And what can be made of the emotional tangle loss presents? Eng and Kazanjian (2003, p. 5) note that "while the twentieth century resounds with catastrophic losses of bodies, spaces, and ideals, psychic and material practices of loss and its remains are productive for history and for politics. Avowals of and attachments to loss can produce a world of remains as a world of new representations and alternative meanings." For this productive hope to be realized, old relations to loss—disavowals, denials, repressions—must be articulated, critiqued and, where necessary, abandoned; new relations with and new meanings of old losses may then be established.

Melancholia and Mourning

Of the expressions of loss outlined earlier in this introduction, melancholia, in particular, offers a rich source of insight into cultural loss and its possibilities for transformation. While each of the structures of loss mentioned here is part of the focus of this book, melancholia is a consistent theme of all of the chapters. Its conceptual importance is reiterated here and throughout emerging scholarship on loss and its politics (Butler, 1997; Cheng, 2000, Eng & Kazanjian, 2003; Gilroy, 2005). While melancholia is discussed in the context of its use in each chapter, an introductory overview is provided here.

In "Mourning and Melancholia," Freud's (1917/1989) seminal essay on loss, he distinguishes between these two grievous responses to loss. He writes that mourning is "the reaction of the loss of a loved person, or to the loss of some abstraction which has taken the place of one, such as one's country, liberty, an ideal, and so on" (p. 586). The psychic process of mourning or "letting go" succeeds when energy is slowly but fully withdrawn from the dead object and the mourner is able to invest in new objects. In melancholia, however, Freud argues that the mourner denies loss, and, instead, reinstates the lost object within the ego in order "to establish an identification of the ego with the abandoned object" (p. 588). Anne Anlin Cheng (2000) notes that "melancholia alludes not to loss per se but to the entangled relationship with loss." (p. 8). Grief is deepened in melancholia because loss is disavowed and not fully accepted and acknowledged. The melancholic ego constituted through such a response to loss is, in a manner, a compilation of its losses, a present identity contingent on its relationship to absents. The melancholic, then, can be reread as one who, rather that living in the past, has an ongoing, living—because one refuses to acknowledge fully death and loss—albeit troubled relationship with the past, with history, with loss. Such devotion to lost objects—or maintaining an illusion of "what was"—prevents a full channeling of this energy into more creative and productive identity formation and to imaging new relationships and new futures. Thus, the melancholic lives in denial of self as the basis for refusing unbearable loss.

Ambivalence, a love-hate relationship with what is lost, is, unsurprisingly, a characteristic effect of melancholia. The melancholic both needs and resents that which is its constitutive base—that which is lost but cannot be acknowledged as lost. Cheng (2000) argues that it is this psychic movement of ambivalence that is the radical basis of melancholia as a critical tool. "[Melancholia] theoretically accounts for the guilt and the denial of guilt" (p. 14). Guilt is a consequence of the ambivalence borne of disavowal. Yet, this guilt cannot be acknowledged, for to do so would be to acknowledge disavowal and, thus, to acknowledge loss. Therefore, as with the denial of grief, the denial of guilt is necessary to maintain the disavowal, the illusion of nonloss. Some of this complex effect is evident in the example of the melancholic migrant, the all-too-common case of the expatriate who struggles to separate from and to confront ambivalence toward a place, a place loved because it is home, yet a place that is also resented because it cannot provide for the migrant a satisfying, safe, or sustainable home. An internalization of negative views of the homeland and repressions and denials of its constitutive base in oneself—a form of self-loathing—are common ways in which such ambivalence is expressed.

Freud initially saw melancholia as a pathological form of mourning in which the mourner was psychically stuck, unable to get over loss and to adequately heal. However, he subsequently revised his initial position and came to see melancholia as a part of the normal process of mourning. Contemporary scholars (Butler, 1997; Cheng, 2000; Eng & Kazanjian, 2003; Eng & Han, 2003), following on and extending Freud's work, critique his initial discrete separation of mourning and melancholia, and, in so doing, they recoup and extend insight into not only the interplay between mourning and melancholia but, more importantly, their coexistence and simultaneity, individually and culturally.

Such contemporary critics see in the melancholia of marginalized social subjects a complex and *necessary* process of negotiation of loss that entails a fierce determination to refuse the loss of precious identities and the social devaluing often attached to them. In this way, culturally, melancholia is a symptom of ongoing assimilation and its necessary losses and, simultaneously, a refusal of these losses. Alternatively, and as well, melancholia can maintain oppressive states, as when it is used to maintain selective national narratives and as means to control and to manipulate the emotional polity. Such melancholic misapprehension is part of an agnotology that is culturally produced, driven by what are often the inseparable interests of nation, military, and media. The same melancholic strategies as those mobilized by nations and states to protect present interests, prohibit critique, and gridlock futures are indicted in all forms of aggression, exclusions, and conflicts. The melancholic's withdrawal to the past can be a form of inattention to the problems of the present and a hostility to what, in the present, threatens one's version of the past. Described in this way, melancholia is a condition premised on aggressive exclusions. It is in confronting this intricate and contradictory dynamic of melancholia that possibility emerges.

Melancholia and Place

A central focus of this book is the vicissitudes of attachment to place and the intricate dynamics of migration as it creates displacement and threatens—as it also recreates—belonging. How might melancholia provide insight into place and the strengths, dangers, complexities, and fragilities of identities constituted around place? Melancholia is the process through which is secured the maintenance of objects lost, abandoned, rejected, threatened, devalued, or delegitimized. In the last half-century alone, Newfoundland and Labrador has struggled with several major cultural shifts, each constituting, and compounding, numerous losses: loss of Responsible Government, nationhood, and independence through Confederation with Canada in 1949; postconfederation government-mandated resettlement of communities in the 1960s; the sexual abuse crimes at the Mount Cashel Roman Catholic Orphanage and other provincial parishes in the 1980s and 1990s; the initial moratorium and subsequent closure of the cod fishery; ongoing postmoratorium resettlement and closure of communities; the unprecedented and, until recently, unrelenting outward migration of the past fifteen years; numerous environmental crises; the increased social alienation of youth and socially and linguistically marginalized communities; the increased social stratification—rural and urban, economically advantaged and disadvantaged—in the wake of oil resource development; and the severe crisis state of many aboriginal communities. The impact of each of these developments reverberates—cast and recast as each is within a potent cultural mixture of love, anger, pride, humiliation, grievance, frustration, and longing—a powerful reminder of how the past is ever present and of how unresolved loss is reenacted, demanding attention and productive address.

In a report of a royal commission formed to examine the place of Newfoundland and Labrador in the Canadian national confederation of provinces and territories—itself borne of the frustrations and limits of social and cultural progress, galvanized by the decline following the collapse of the cod fishery—it has been argued that, within the province, these developments have not been fully confronted nor productively processed for the losses they represent (*Our Place in Canada*, 2004). Nor have the psychic, social, and cultural relations to these multiple losses been adequately examined. Melancholic attachments are evident on multiple social sites: a preoccupation with the past in art and culture; the struggle to accept the threatened condition of the cod stocks through, for example, an insistence on a recreational food fishery; the persistence of the language of death and dying to describe negative cultural change; and the reiteration of old views and old patterns that not only fail to meet the demands of new circumstances but also inhibit change. Indeed, it is reasonable, in relation to certain persistent positive cultural stereotypes of character—imposed from without and within—to ask to what extent the cultural expectation to be hearty and resilient might pose some kind of cultural prohibition on full grieving. In addition, it is important to consider to what extent the compulsion for humor might be, among other things, also a repudiation of mourning.

Such numerous disavowals shore up the complexities of a melancholic subject, one who, depending on social position— for example, the White setter in relation to aboriginal peoples— can be seen as both colonizer and colonized, disavower and disavowed, and who, in refusing to grieve or to acknowledge that grieving is a necessary part of renewal and change, remains culturally caught "between hope and despair" (Simon, Rosenberg, & Eppert, 2000). In the absence of a public space in which to examine disavowal—that is, a space in which *to avow loss*—a refusal to grieve or sanctions on grieving compound the crisis of losses, the disarray of identities, and the social and cultural dysfunctions that are its consequence. Creating conditions for the avowal of losses, beginning and supporting the process of mourning, creates the possibility for an articulation of other denials and disavowals. Only then is it possible to engage the disavowals. For example, despite a history of genocide—the Beothuck, a group of aboriginals who inhabited the place prior to European settlement and who were eliminated through slaughter and disease by the early nineteenth century—and current disaffection and suffering among aboriginal peoples in the province, it is not uncommon to hear the claim that racism is nonexistent in Newfoundland and Labrador due to the perceived sameness of the population.[8] This particular disavowal (of any number of differences, of which racialization is one) is an historical effect of colonization, and its reiteration is a contribution to conditions that do not preclude discrimination but that can, instead, create a seedbed for its continuance and growth. Dionne Brand (2001) comments on similar denials of the "I was not here, I did not do that and feel that" sort, which resist implication and responsibility: "One hears that all the time in Canada; about what people feel they are or are not responsible for. People use these arguments as reasons for not doing what is right or just. It never occurs to them that they live on the cumulative hurt of others. They want to start the clock of social justice only when they arrived. But one is born into history, one isn't born into a void" (Brand, 2001, pp. 81–82). An analysis of the sort of implication to which Brand refers is part of the work of confronting loss and recognizing the dangers, as well as the possibilities, of melancholia.

Melancholia and Reparation

The complexities of identity, and what Megan Boler calls their "precariousness and frailty" (Boler, 1998, p. 194), to which melancholia in part attests, is what makes change—and education for change—both possible and difficult. The very arbitrary nature of social positioning reveals the fundamental vulnerability of us all. Jonathan Lear (2006) writes of an "*ontological vulnerability* that affects us all insofar as we are human" (p. 50):

> If there is a genuine possibility of happenings' breaking down, it is one with which we all live. We are familiar with the thought that as human creatures we are by nature vulnerable: to bodily injury, disease, ageing, death—and all sorts of insults from environment. But the vulnerability we are concerned with here is of a different order. We seem to acquire it as a result of the fact that we essentially inhabit a way of life. Humans are by nature cultural animals: we necessarily inhabit a way of life that is expressed in a culture. But our way of life—whatever it is—is vulnerable in various ways. And we, as participants in that way of life, thereby inherit a vulnerability. Should that way of life break down, that is *our* problem. (Lear, 2006, p. 6)

What might it mean to accept this condition of vulnerability, to acknowledge this fundamental fragility in a wide-awake and conscious way, not as a mere statement of fact but as an ethic of living—to fully accept such vulnerability as "our problem" and about which something must be done on an ongoing basis, and not only when catastrophe looms or a crisis is pending? In the Report of the Royal Commission on Renewing and Strengthening Our Place in Canada, the commissioners spoke to this fragility as it relates specifically to Newfoundland and Labrador, stating that as a result of historic losses (of governance, nation, and independence), "our identity and sense of place are, and perhaps have always been, vulnerable" (*Our Place in Canada*, 2003, p. vi). Such vulnerability is often expressed in a "bunker mentality," identifying oneself as under siege and holding fast through hypervigilance and overwrought responses to criticisms. Rex Murphy, commenting specifically on Newfoundland and Labrador, referred to such cultural vehemence as "a stand-in for something [that] speaks to the uncertainty of something really central. I'm speaking in the long-term and survival of the culture. Is this province as we know it...going to maintain itself?" (Porter, 2005).

However, this stand-in, a reiteration of historic and contemporary disavowals, is not a productive means of cultural maintenance or betterment—to the contrary, for its primary impetus is regressive and narrow. A powerful desire for a partial affirming knowledge—stories that assuage fragile identities but that are also based on trenchant disavowal, for example, that foreign overfishing is at the heart of the collapse of the cod fishery—can override other possibilities. There may be more to be gained in facing loss and our own implication in its circumstances, as difficult as that may be. Judith Butler (2004) poses "an apprehension of a common human vulnerability" based on a consideration of the vulnerability of others through a mourning that enacts "the slow process by which we develop a point of identification with suffering itself" (p. 30) as the basis for a new politics of difference. Ing Ang refers to the need for "a pragmatic faith in the capacity of cultural identities to change, not through the imposition of some grandiose vision for the future, but slowly and unsensationally" (p. 11). Yet, such unfolding, as Butler suggests, can be understood not as happenstance but as a process of mourning borne of the recognition that what is happening requires our openness, inventiveness and courage. Up against a tidal wave of forces with global dimensions, a place cannot and should not attempt to maintain itself through regressive narratives of place and identity, although such is what commonly happens.[9] Change necessitates as much attention to the disorders of identity, its melancholic afflictions, for example, as well as to its mythic ideals.

Melancholia offers as its most potent positive thrust toward injury and vulnerability the urge for reparation. Melanie Klein differs from Freud in her account of the psychic processes of mourning. She contends that any mourning (of a loved one, an ideal, a place, etc.) reactivates early psychic processes begun as a response to conflicted feelings (of love and caring and anger and resentment) created by the loss of early idealized loved objects (usually the child's caregiver). She argues that from these conflicted feelings, that is, ambivalence, arises guilt, and from this guilt comes the desire *to make reparation*, to repair or to make good on the injuries done to the loved object through such vicissitudes. Klein argues, through the urge of reparation, largely gained through experiences of loving and caring, a foundation for security, trust, love, and a belief in goodness is established. Reparative processes and practices continually integrate the psychic and social worlds. Klein argues that residues of these early psychic dramas accompany us throughout our lives and are replayed as new editions of old conflicts in our ongoing struggles to contend with loss and to reestablish security and love in our worlds. In this sense of living in uncertainty and loss, then, reparation, as a means to recreate a lost good object, gets its impetus from our most productive and most creative human inclinations.

What might an understanding of reparation enhance an understanding of the melancholic subject? Rinaldo Walcott, writing about the African diaspora, uses Klein's notion of reparation in a discussion of a creolized pedagogy. For Walcott, a creolized pedagogy is one in which the complexities, ambivalences, and (re)inventiveness of identity might be recognized, analyzed, and interrogated. Walcott contends that "creolization offers a way of seeing defeat as more than lost/loss. It is a practice that may impede the transformation of defeat into (our) shame" (Walcott, 2000, p. 139). Walcott suggests that the injuries of history and the renegotiations of identity thus incurred are not sources of defeat or shame. Instead, he argues that reparation is central to this work of confronting productively the presence of repeated trauma in relation to one's identity. That is, confronting ambivalence, the love-hate relationship of the melancholic "is [a way of] making reparation to ourselves, an acceptance of the ways in which the pain and the pleasures of history have intervened to invent us" (p. 149). Just as movement has been part of the processes out of which present identities and attachments have been formed, so, too, will new ones arise with changed circumstances—the vicissitudes of history. Thus, we are not disloyal if we move away and change; we are not abandoning our parents and ancestors if we leave; we are not less if we stay; and, we are not more because we go. Walcott emphasizes, in concert with Klein, the importance of collective, critical self-love and self-acceptance (of a self that is fragmented, creolized, and hybridized) as a precondition of ethical relations with others.

Toward an Educational Discourse of Loss and Place

In such a pedagogy as the one described by Walcott, problematic attachments can be rethought and renegotiated in a space created to offer a reparative gesture to oneself. In this space is also created the opportunity to form new relationships to old attachments, relationships that bear the mark of responsible engagement for change. But for such work to begin to be successful, lost and displaced objects must be clearly scrutinized for what they tell us about how our structured responses to loss have both helped us *and* hindered us. A haunting allegiance to place (and its melancholic ghosts) requires more intricate analysis and greater examination of its regressive tendencies and its possible sources as psychic injury. In the case of Newfoundland and Labrador, for example, it is not possible to separate such allegiance from its place within a broader history of colonization—and the historic disavowals that brought immigrants (largely from Ireland and England) to this place and the current disavowals that are its legacy and a part of an (ongoing) taking of the place. Crucial to such an examination is a wrestling with the complex "psychic citizenship" (Eng & Han, 2003, p. 266) of our first peoples, early and later immigrants, and those who carry forward and replay these psychic legacies through the various social and cultural struggles that constitute what it means to stay in or to leave the place.

Such issues—of melancholy, ambivalence, and reparation—are central to an examination and any discussion of what it means to educate across social and cultural difference in the context of loss. Such an education would confront and elicit a range of emotions, many of which can be identified as symptoms of latent melancholia, a confirmation of the presence of disavowed others who constitute our beings and who haunt our present. Yet, to understand melancholia in its complexity is to see these feelings not as emotional obstacles to avoid but as critical opportunities; that is, to see these feelings as both regressive and potentially productive, and as the basis for building reparative relationships with disavowed others.

The allure of the myth of redemption—recovery, forgiveness, and progress, a longed-for sense that the future will somehow make the past "worthwhile"—is ever present: We will be free, independent and economically respectable; the fish will return; those who left will come home. Learning must foster hope, but hope with a critical base, that is, a critical hope that, as Boler suggests, "directly challenges inscribed habits of emotional attention and signifies a willingness to exist within ambiguity and uncertainty" (2004, p. 129). Such a hope is premised against the dominant and habitual urges of melancholic to entrench through disavowal and to rebuke change. It ushers in a learning more precarious and painful, but also more hopeful and productive. Simon, Rosenberg, and Eppert (2000) argue for such learning as "a learning to live with loss...and to live, not *in* the past, but *in relation* to the past" (p. 4), a learning "to live with what cannot be redeemed, which must remain a psychic and social wound that bleeds" (p. 5). Here, the necessary work of mourning in learning is evoked. So, too, is the specter of the melancholic, in whose unfinished work of mourning lies not just the evidence of a disappointed past but also the impetus for a critical project of loss, contained within a hope of a future that ethically carries forth the past as always with us.

Finally, Judith Butler (2003) encourages us to see the possibility in this animation that is loss. She reminds us, "Places are lost—destroyed, vacated, barred—but then there is some new place, and it is not the first, never can be the first. And so there is an impossibility housed at the site of this new place. What is new, newness itself, is founded upon the loss of original place and thus...[is] fundamentally determined by a past that continues to inform it...an animating absence in the presence" (p. 468).

Butler suggests that a new or renewed place, rather than being a place of "no belonging, where subjectivity becomes tethered from its collective fabric," is instead "a place where belonging now takes place in and through a common sense of loss....Loss becomes condition and necessity for a certain sense of community" (p. 468). Such an education for social and cultural difference in the context of loss must be committed to a learning that encounters the place of loss in place—an educational discourse of loss that is an active part of considerations of change. Such an education would encourage new forms of belonging, something that may only be possible or feasible when older, more melancholic forms of belonging are explored, understood, and contested.

The Book

The essays in this book address these broad issues of culture, loss, and identity, in particular, as they relate to themes of migration, melancholy, and reparation, and how, together, they intersect with and have implications for an educational discourse of loss and place. Through these lenses, the essays focus on the emerging context of Newfoundland and Labrador at the beginning of the twenty-first century. In so doing, they individually and collectively inquire into the structures of feeling which constitute loss. Loss—its tangled vicissitudes and effects and the complex processes by which it is registered and processed—is not only individual but communal, and it is a matter not only of the psyche but of society, culture, and politics, also. These essays delve into the tensions that accompany personal and cultural loss and how politics and culture shape them all.

While each chapter of the book is discrete and stands alone as an essay, all of the chapters are thematically interrelated. Each chapter deals, in different ways, with the central questions: What does it mean to speak of cultural loss and mourning? What does it mean to investigate cultural loss as a political and psychic process? How do personal and cultural losses intersect and coalesce? How might such issues of loss be engaged by educators? As an amalgam of autobiography, theory, analysis, and critique, each chapter, to a greater or lesser degree, also attempts to traverse spaces too often separated in academic writing. To enhance this bridging of theory and autobiography and emotion and intellect, short fictional vignettes, in the form of a *postscriptum* for each chapter, are used as a different space through which to prompt a further and different (re)consideration of the thematic foci of each chapter.

The opening chapter, "Losing Place: Reluctant Leavings and Ambivalent Returns," explores some of the dimensions of a relationship of personal and cultural migration and loss. In locating my own story of leaving (and returning) within the framework of a culture of largely economic migration, my purpose in this chapter is to raise questions not just about migration and transmigration but, perhaps more so, about the cultural impact of collective loss and the challenges and opportunities it offers for change and renewal, through mourning. With a focus on education, I attempt to trouble the cultural and institutionalized silences around loss, death, and grieving as a means to understand the cultural prohibitions on loss and their costs. In so doing, I address what loss and grief might demand of us by way of social, cultural and educational responsibilities and opportunities for insight and change. In this sense, the chapter points, more generally, to hopeful possibilities that may be accrued individually, collectively, and educationally from *a reflective grief*.

The subsequent chapter, Chapter 2, "'The Distance Home': Migration, Mourning, and Difference," continues the examination of the vicissitudes of leaving and return, and of reconciliation and renewal in the face of loss. This examination is based in an analysis of Lawrence O'Toole's memoir, *Heart's Longing: Newfoundland, New York, and the Distance Home*, which recounts a life journey to find "a place of the heart." The memoir oscillates between two home "villages" in crisis—the author's home in the "gay village" within Manhattan at the height of the AIDS epidemic and his childhood home in the fishing village of Renews, Newfoundland, following the closure of the cod fishery. The memoir is presented as an example of writing as reparative and as a study in how sexuality as difference complicates issues of loss, longing, place, and belonging. This chapter furthers the importance of finding and articulating grief as an integral part of personal, cultural, and educational change.

Chapter 3, "The Word, for Loss: Literacy, Longing, and Belonging," problematizes attachment, place, longing, and leaving through an exploration of the psychosocial aspects of literacy and literacy education. Focusing on the *structures of feeling* (Williams, 1973) that might accompany some prevailing and persistent beliefs about literacy, historically and in the present, the chapter highlights the complex intersections of history, memory, culture, and identity, and how their relational character may come to bear on longings for, beliefs about, and cultural practices of literacy. Framed around a discussion of the historical verse novel *Ann and Seamus*, written by Kevin Major and illustrated by David Blackwood, this chapter suggests some of what psychoanalytic inquiry might offer literacy research, in particular, and educational studies, more generally. The discussion inquires into the nature of certain attachments and their intersection with projects of literacy as they relate to an education for social and cultural transformation.

Chapter 4, "Separation, (Re)connection, and a Transformative Education of Place," addresses the educational implications of the ecological and environmental challenges that disrupt and reshape attachment to and sustainability within a place. Blending educational narrative and cultural analysis, it inquires into particular structures of attachment—nostalgia, solastagia, and melancholia—as symptoms of identities in the throes of upheaval and change. It also offers insight into forms of attachment that might promote self and community efficacy and forms of education that might address the dilemmas of environmental loss and degradation and enhance possibilities for growth, renewal, and change.

Chapter 5, "The Place of Reparation: Loss, Ambivalence, and Teaching," investigates the seemingly banal nexus of biography, geography, and pedagogy as psychic passions, as constituted (and constitutive) sites of ambivalence in teaching and learning. Specifically, I strive to establish a series of connections among telltale passions, beginning with an investigation of what might lie beneath a lifelong set of loyalties, beliefs, and practices. In pursuing this series of connections, I focus, first, on the psychoanalytic notion of reparation, as developed by Melanie Klein, and its usefulness in an analysis of autobiography and teaching, specifically, and teaching, more generally. I inquire into a desire to teach by asking on what impulses such a desire is based and what (reparative) urges might constitute a love of teaching. Secondly, I explore the relationship of a love of place and a love of teaching in relation to reparation and teaching. I conclude with a discussion of teaching as reparative within a notion of teaching rendered overtly political, the desires of which reach toward justice and education across social and cultural difference.

The concluding essay, "'Learning to Live with Ghosts': Loss, Place, and Education," provides an overview of an education that centers loss. It outlines the key characteristics of an emerging educational discourse of loss and place. I have argued throughout these chapters that cultural crises demand a rearticulation of educational vision at local and global levels, as is well demonstrated in the work of many contemporary scholars. In each of the preceding chapters, I have addressed aspects of a cultural crisis particular—but not limited—to Newfoundland and Labrador, and some of its ecological, social, personal and psychic dimensions. In so doing, I have gestured to a notion of education in such a context that challenges accepted traditions within mainstream and, as well but to a lesser extent, critical practices. In this concluding chapter, while drawing from these preceding discussions and returning to issues outlined in the introduction, I attempt to coalesce what I consider to be key characteristics of an education that addresses cultural crisis, loss, and change, more generally, and in the context of a multicultural world.

There are many who might argue with the premise of a book that focuses on issues of cultural loss in relation to Newfoundland and Labrador at a time when a new sense of economic optimism is emerging, as a result, in large part, of increased oil revenues gained against a backdrop of a world energy crisis. The newfound optimism, while welcomed, will not override the deep cultural problems to which this book speaks. Nor can or should economic advancement be used as an excuse to ignore the persistent patterns born of disaffection and injury, incurred and levied. Economic upswings can be short-lived—and the forecasts for the province suggest that this will be the case here. The challenge, now, is to seize this time of greater economic certainty to forge a new cultural and ecological ethic that will sustain a different kind of future, one in which we are better positioned to address the inevitable losses that will continue to mark us, as a culture. It remains to be seen to what extent this may be achieved. In this regard, this current book, then, like the first, *Marketing Place*, ends on another threshold, as a culture prepares to contend with what a new economic reality might bring its way.

CHAPTER 1

LOSING PLACE
RELUCTANT LEAVINGS
AND AMBIVALENT RETURNS

Emigration, forced or chosen, across national frontiers or from village to metropolis, is the quintessential experience of our time.

J. Berger, *Keeping a Rendezvous*

Storytelling provides a resource for the constitution of identities and an explanatory framework for particular experiences.... [It] both enables and embeds the performative construction of collective and individuated identities; it provides a structuring dialectic mediating the imaginary and the material.

A. Kear and D. L. Steinberg[1]

INTRODUCTION

The migrant story, lived and told thousands upon thousands of times in this past century of massive worldwide migration, is a story that holds the peculiar tensions of loss and hope as its central premise. At the least, it is a story of physical displacement, one through which a citizen or subject abandons, under compulsion or voluntarily, a place of habitation and the cultural practices of community that constitute it. More profoundly, it is a story of affect, a severing that disrupts and reconstitutes one's sense of self and community, triggering a plethora of emotional and psychic struggles. Recognizing the number and extent of these losses and the impossibility of speaking of loss without also speaking of mourning, Eng and Han claim that "the experience of immigration itself is based on a structure of mourning" (2000, p. 352). In this chapter, I explore some of the dimensions of this relationship of migration and loss, from both a personal and a cultural perspective. In locating my own story of leaving (and returning) within the framework of a culture of migration, my purpose is to raise questions not just about migration but, perhaps more so, about the cultural impact of collective loss and the challenges and opportunities it offers for change and renewal through mourning.

Within the social sciences, and education in particular, questions of how to educate in the context of profound cultural loss and how to acknowledge and work with the vicissitudes of mourning are still very much underexplored (Kessler, 2004; O'Sullivan, 2002). Ruth Behar (1996) offers some explanation for the dearth of attention to such issues within academia, more generally. In writing about her work as an anthropologist—to analyze and to document the dying cultures of "others"—she argues powerfully for the need to speak differently and more fully about "ordinary death" and how we die. Following de Certeau (1984), she refers to contemporary practices around death, which she argues effect a distancing of death and the dying, a rendering of it as "other." Behar argues that such distancing practices have a material base, ensuring "that [death] will not interrupt the chain of production and consumption that keeps our capitalist cultures running" (p. 85). Death, it seems, cannot be allowed to disrupt the economies of efficiency that organize contemporary (postcapitalist) life. Behar's point is realized in the well worn adages that are underpinned by an ill-conceived notion of closure following loss: life must go on; there is work to do; time is money. As such, these too-rarely questioned beliefs not only protect a certain materialist ideology, they constitute a particular emotional economy and, as well, one in which such adages, along with a collective discomfort with death and mourning, are acritically accepted.

Behar further argues that, within academia, it is "the fenced boundary between emotion and intellect that is the academic counterpart to the practices that make dying 'obscene'" (p. 86). This false yet firmly established division of emotion and intellect is well known and, while its politics have been articulated and challenged by many scholars, there are more than residual remains of such Enlightenment thinking (Kincheloe, 2000). In this chapter, I want to focus on some of this scholarship that resists these old divisions, investigating the place of loss in contemporary social formations, thereby providing new spaces for reconsidering intersections of the personal, social and cultural in relation to "grief work"—the emotional and psychic work of dealing with, coming to understand, and living within the spaces created by loss and mourning. With a focus on the impact of migration, particularly outward migration, as an experience of loss, I am interested in troubling the silence around loss, death, and grieving in the academy; perhaps more importantly, I want to address what loss and grief might demand of us by way of social, cultural, and educational responsibilities and, as well, what it might offer us by way of opportunities for insight and change. In this sense, the chapter points, more generally, to hopeful possibilities accrued individually and collectively from what I am calling *a reflective grief* prompted by, although not limited to, experiences of migration.

Loss, Grief, and Identity

Grief is more than a privately constituted emotion. While grief has deeply private, even esoteric, dimensions, the meanings and practices of grief—how we grieve, how we understand loss and death, how we understand what we may claim as an experience of loss, and what we feel is or is not permissible in mourning—are socially embedded. Every culture has its rituals and meanings around loss and mourning and, while these may change over time, they remain heavily normative, tied to dominant, albeit not always stable, prevailing notions of the work of grieving. Through normative social practices, then, this work of grieving can be said to be "managed" or, as Richard Johnson (1997) argues, "patrolled." Johnson writes,

> [I]t seems to me that death is another kind of "border" and that grieving and its accompanying activities are another kind of "patrol." Like all "border patrols," grieving rituals and practices are there to police the boundaries....Mourning is a work of reassurance and boundary—maintenance against the shock of death and loss. But the reassurance is necessary—even though it many not succeed—because the death of someone close to us produces a "madness": overwhelming emotions that throw into giddy, vulnerable, high relief *all* our own identities. (Johnson, 1997, pp. 234, 235)

Here, Johnson points to the necessity—and, oftentimes, the failure—of mourning, and, while his comments come out of a sense of personal loss, they carry over into a notion of collective mourning. Cultural loss also produces a "madness," one made more difficult because, in many contexts, our understandings of and rituals for cultural mourning are muted and ill-understood—and the sanctions against it, as Behar noted, are so pronounced. Judith Butler (2004) comes closer to framing the sociality, the relationality, of loss:

> When we lose certain people, or when we are dispossessed from a place, or a community, we may simply feel that we are undergoing something temporary, that mourning will be over and some restoration of prior order will be achieved. But maybe when we undergo what we do, something about who we are is revealed, something that delineates the ties we have to others, that shows us that these ties constitute what we are....When we lose some of these ties by which we are constituted, we do not know who we are or what to do. (p. 22)

Understood this way, loss and dislocation can be viewed as an assault on identity, a challenge to the very constitution of who we are and who we see ourselves to be. Such insights readily apply to aboriginal peoples, such as the Innu of Labrador, and their suffering and loss effected through forced relocation (Ellwood, 1996). These insights also relate to other migrations and the losses they summon, both for those who leave and those who are left, as both are changed through the mutual loss migration creates. The seemingly mundane and ordinary stories of migration reveal the profundity and power of this experience.

What happens in the borderlands of grief, in these in-between spaces of "madness" wherein the "boundaries" of identity are pushed and confronted, where we become, in ways, "undone"? Inhabiting these spaces as part of my own griefwork, I want to trace how various losses have disrupted my own cultural identity, in particular, my sense of place, my *Newfoundlandness*. As well, through addressing the implications of the place of issues of loss, death, and mourning in academia, I also pose questions about my professional identity, as an academic, and what this disruption of identity produced, in terms of how I understand what it means to educate in a context of cultural loss and mourning, to educate in and for cultures afflicted by loss. The Canadian poet Lorna Crozier (1999) writes that "death leaves us our afflictions—they're what the living won't let go" (p. 18). But if loss and death bequeath what some may consider burdens, they are burdens of responsibility: to remember, to reflect, to recompose. Such responsibilities can be explored in the spaces created by loss and grieving that I argue can be productive spaces for renewal—rich educational opportunities through which to explore the many conflicting and contradictory dimensions of loss.

MIGRATION:
A STORY IN THREE PARTS

There are many ways to tell a story. In opposition to. In sympathy with. What to leave out. What to put in.

H. Humphreys, *The Lost Garden*

Sometimes I think that if it were possible to tell a story enough to make the hurt ease up, to make the words slide down my arms and away from me like water, I would tell that story a thousand times.

Shreve, *The Weight of Water*

In what follows, I trace some of my own experiences of leaving as a means to ask questions about cultural loss in relation to these experiences. While framed by the privileges of social class and position, my story still for all forms what I believe is a viable basis for an examination—but by no means a universalization—of issues of identity, performativity, and (be)longing as they relate to place and profession and as they are unsettled and transformed through loss. In this story in three parts—entitled "Diaspora," "Curriculum Vitae," and "Nostalgia," respectively—I attempt to locate the individual against a cultural backdrop, shunting back and forth between the coincidences of loss and change as they unfold on these two sites as a means to draw connections between personal loss and what Rachel Kessler (2004) calls "tribal losses" (p. 150), losses that affect an entire community, culture or nation.

1. Diaspora [Gk. *dia*, apart; *speirein*, to scatter, sow]

Political exile or economic displacement and the eradication or "death" of communities and nations and the ways of life they support, as well as the disruptions to cultural identities of such patterns of movement create deep cultural marks. After Butler and Johnson, it can be argued that such changes place a premium on the renegotiation and redefinition of communities and identities. In this sense, migration and diaspora are sites of troubled hope and potentialities. Through both migration and diaspora, the home place and its affiliates (or expatriates) can be challenged and changed. In relation to diaspora, Braziel and Mannur (2003) note, "Diasporic traversals question the rigidities of identity itself—religious, ethnic, gendered, national; yet this diasporic movement marks…a nomadic turn in which the very parameters of specific historical moments are embodied and—as diaspora itself suggests—are scattered and regrouped into new points of becoming (p. 3)."

Both migration and diaspora can challenge the established "certainties" of home place, encouraging an examination of experience—old and new—that produces new knowledge and additional layers of insight. These insights can produce new forms of identity and meaning, sometimes revealing previously eclipsed or naturalized differences as noteworthy and productive.

My own cultural identity wrests in parental lineage of both Ireland, through my father, and England, through my mother. This conflated and conflicted duality resonated in early learnings and more forcefully so as I grew older and learned more about the travesties of colony and empire in relation to Ireland

and Britain and Britain and Newfoundland. This ancestral split, merged through various migrations, registers still as a faint haunting, a sense that parts of my self originated elsewhere and nudge still, interminably, for reconciliation. However, in my social studies courses, the presence here of the Irish and the British was presented largely as natural and entitled, although differently for each. At school, we studied nomadic peoples in geography, waves of migration in history, and the victories of empire everywhere, but we did not consider, through any of those studies, how people felt when they moved—or were forced to settle, which was the case for so many aboriginal peoples—or what demands and difficulties such movement evoked for those leaving and left behind and for those others on whose shores they would arrive. And we were certainly not taught multiple perspectives on these migrations. Movement was one in a long series of givens—experiential and educational *faits accomplis*. In later school years, community professionals (teachers and doctors) representing newer diasporas within the province (immigrants from Sri Lanka, the Republic of China, the Philippines, India, and Pakistan) along with other "newcomers" from the United States, and mainland Canada, provided rich opportunities for personal and community growth. European Jews, some survivors of the Holocaust, built thriving businesses in a nearby community.

I know the pain of difference marked those times, but also the joy of mutual renewal and gain—the hopeful dimensions of a diaspora's sowing and scattering. But I suspect that, despite the immense contributions of these gifted people to my life, my small community, and the province, few would have been seen—or, perhaps, would have considered themselves—to be Newfoundlanders. And, I fear that dislocation and loss must have haunted each of them in their own way. I did not begin to fathom these ruptures until I experienced some version of them myself, in another context in which the stories of migration and exile circulated widely, having vividly marked so many who would become friends. Caring encounters form the basis of ethical relations of the sort on which healthy communities can be built. The myth of "sameness" which still prevails in Newfoundland and Labrador is an effect of discourse and power that is undermined by my own account of a childhood immeasurably and positively shaped by the gifts of others who, despite the limits placed on their belonging, converted their losses into my and our gain. Could there have been a place for such concerns—of migration and identity—in those classroom spaces we inhabited so long ago? Should there be such a place now?

The 2006 Canadian census documents an 11.08 percent decrease in the population of Newfoundland and Labrador—approximately 63,000 people—since 1991 (Antle, 2007), the eve of the cod fishery moratorium that halted the economic lifeblood of hundred of communities. For a province with a long history of economic migration (Overton, 1996), stories of leaving are an indelible part of the collective cultural psyche. This story of leaving or migrating—or, in colloquial terms, going "to the mainland" or "upalong" or "out West"—is reiterated on many fronts as "a story," a specific story of "someone," while, at the same time, it is linked to a recognizable cultural story, one which is shared by many, and is an established and well-known feature of a particular culture. Both in its technical and emotional structure, and whether told in art, prose, verse, or song, the migration story nears the status of genre, with its telltale conventions and features: leaving one's home and family, often out of (economic or political) necessity, to work and live elsewhere; the contradictions of displacement and alienation felt in "being away" alongside the realities of a more physically or materially secure life; the consequences of family fragmentation; the conflicted longing for "home"; and, usually, some reconciliation of the impossibility of return. Such a story is not limited to the experiences of the migrants of Newfoundland and Labrador. The twentieth century has seen more migration globally—voluntary or forced—than any other. The vast literature of migration and exile documents the social, cultural, and personal consequences of these global trends of displacement and change.[2] Still, the migrant story in relation to Newfoundland and Labrador—or any other geographical place—has its own particularity.

Kear and Steinberg (1999) argue that "stories provide an ephemeral space, which is at the same time an organizing *performative* matrix for the regulation and deployment of material power" (p. 9). The relationship of the romance story to compulsory heterosexuality and the adventure story to imperialism are but two examples that demonstrate this point. The formulaic structure of each of these genres naturalizes the emotional and political economies out of which each arises as a story. In previous writing about the migration patterns of Newfoundland and Labrador (Kelly 1993), I argued that the emotional economy of migration (displacement and nostalgia, in particular) secured the divisions and disparities of the Canadian national economy: a surplus army of workers on site, when needed, at home (in the "regions"), when not. While this economic analysis, in the context of the rampant globalization of the past decade, now needs modification, the migration story remains a performative part of normalizing the inequitable social relations that form the basis of a transnationalist economics of and in global capitalism (Braziel & Mannur, 2003). This global economics is marked by intensification of urbanity, accompanied by a massive decline in rurality, the results, in large part, of the necessary movements of labor in response to the demands and ravages of these economies—trends also borne out by the 2006 Canadian census *(The Globe and Mail,* March 14, 2007).

In any discussion of migration, it is important to avoid any inclination to collapse or homogenize all migrations and instead to note the specificity of context—political, economic, affective—out of which various diasporas have been or are created. However, an analysis of migration as it pertains to largely White settler colonies can highlight "the complexities of the settler subject, who *as both colonizer and colonized* [italics mine], occupies a uniquely ambivalent position" (Whitlock, 1994, p.1). In the history of nations and provinces, Newfoundland and Labrador is decidedly young; in the history of colonies and empires, it is not. Its long history of colonizing and colonized identity formation is registered in and through present cultural forms and discursive practices. This long history of migration, and, in particular, this most recent "wave" of outward migration, brings several questions to the fore: What happens when these historical White "settlers" are dispersed, resettled? What do the stories of the economic migration of settlers from their (former) "colonies" reveal? What is the social, cultural, and emotional impact of these migrations for the place of leaving? What educational issues might arise from an examination of such questions?

2. Curriculum Vitae [L. *currere*, race, course; *vita*, of life]

I had left Newfoundland, initially to study, in the early 1980s, at a time when the population of the province was nearing its peak, at around 568,000 people. Wrestling with the impact of leaving and the struggle to integrate "others" and "elsewheres" complemented my work in critical education, which provided me both an analytic lens and a discursive space to examine the nature and complexity of my experiences. However, at the time, it eluded me that loss was the central feature of what I was experiencing through heightened feelings of difference and displacement. It was not that I did not experience loss or the longings it fuels; it was that I experienced its effect more as *longing* than loss. When I did experience loss in the galvanizing and riveting manner that can be the impetus for profound learning, the opportunity to reflect deeply on the implications of that loss was hard to find in time preoccupied with the demands of work. A seemingly small moment would change that.

In the summer of 1989, while attending an international institute in teaching and learning, on another island not home, I was among participants who were asked to construct a lifeline of significant events, a form of *curriculum vitae* academics are rarely asked to compose. At the time, two events—each of profound displacement and disorientation—dominated my account: leaving Newfoundland to study and, subsequently, to work—a cementing of the leaving, and the death of my father during my first year of full-time academic work. These two events were inseparable then; they feel inseparable still. In my scripting of these two events into a relationship, leaving became synonymous with loss. Nine years later, in 1998, while I was still away, my mother died. Two-and-a-half years following her death, I decided to return to Newfoundland. These events became inseparable, too. If leaving was about loss, returning was about recovering (from) loss. While I was away, (pre)occupied with the academic (race) *course of life*, both my parents died: my time "away" was a *curriculum vitae* bookended by loss. These events are not permissible entries for any academic vita, yet the attachments that render them meaningful form the heartbeat of the work I do and shape the style and manner in which I do it. What did this mean? What might I learn from an examination of these realizations, this relationality of loss?

The minutiae of our lives are saturated with meaning, and stories are an opportunity to mine this minutiae, to forge connection and insight. In this sense, minutiae feels monumental. Here, then, is minutiae as monument:

> During his life, my father had refused to leave Newfoundland, even to vacation with his family (who went without him). He would say, "Why go elsewhere when you are already in the best place there is?" Perhaps he had just had enough of moving about, house and all, having "resettled" for economic reasons twice before finding a permanent home where my siblings and I grew up. My mother, on the other hand, left Newfoundland often, usually to visit or to travel with my brother and me. Although no less a proud Newfoundlander than my father, she had raised her family to succeed, to get out (at least of the community in which we were raised). Outward mobility was upward mobility, her desire for all her children. A strong, proud woman of immense dignity, in keeping with the expectations of the time, she reluctantly relinquished her career as a teacher to marry and to raise a family. I continued her dream. My accomplishments, I reassured myself, were some antidote to her regret, her loss. Over those years I was away, outside of all too infrequent visits, our points of contact were by phone and mail, those difficult connecting spaces for a mother and daughter estranged by migration. I feared her loneliness and she mine. She, after all, understood the loneliness of leaving better than I. She, not unlike me, but unlike her other siblings, had left her hometown, to teach in various small communities in Newfoundland in the dire years of Commission Government, prior to Confederation with Canada in 1949. Ironically, her death came suddenly, while walking to the post office in my hometown—seeking connection to the very end. It was a Monday morning, for me the beginning of a work week, another week of longing. I was at work, doing her work—teaching—when I received the news of her death. That morning, I walked out of my office, catapulted ahead in that longest journey—toward self, meaning, connection—by this unbearable loss. Each individual death comes down to a heart which stops beating. The death of my parents: hearts finally "broken." My

return to Newfoundland: postmortem of a grievous heart attempting to reconcile the losses and incongruities created through migration.

As is the case for most Newfoundlanders and Labradorians, living away presented vivid opportunities for me to confront how our provincial identity is constructed nationally, as "other" (although it is not necessary to leave to experience this "othering"). Like Susan Tilley, through many of my experiences, I came "to recognize the depth to which I was personally marked by culture, history, and language" (Tilley, 2000, p. 241) as I lived daily incidents that demonstrated a broader devaluing of many of the things I had been taught, at home, to love and to honor. The halls of academia are not the same as the floors of the factory, but the dynamics are similar, as neither space exists outside of the social and cultural. Being middle class only shelters somewhat. These experiences—of confronting "the stories others tell of us"—galvanized what was already a strong claim to place. I began to introduce myself, first and foremost, as a Newfoundlander, by way of dare, proclamation, pride. I began to "perform" my place identity in ways that were both productive and reproductive, (re)membering and reinventing for myself and others "who I was," claiming my difference but refusing how others often constituted it as pejorative, a source of ridicule—disparity—being "that," but "not that," both more than and less than. In the meantime, I taught and learned and became a part of the context of others' histories, struggles, cultures, transgressing difference through shared experiences of loss and alienation—strong bases for coalition building.

Part of the chaos and disorder—what Richard Johnson earlier deemed *madness*—that, for me, accompanied my mother's death, in particular, was a questioning of the everyday practices of the academy and the culture of which they are a part. Being away "to work" can come to mean that work is the central defining feature of one's life, a meaning reiterated through the demands and desires constituted through contemporary forms of materialism and rampant consumerism. When one is without family, and with friends often defined through and around work, an infrastructure of emotional support can be limited and ahistorical. Differences that can usually be successfully negotiated can be thrown into relief by excessive trauma or loss that leaves one grieving outside the cultural context of the loss. Within the academy, the antiseptic silences or muted acknowledgements that accompany death, which refuse to name it or note how it reconstitutes bodies, fiercely reiterate boundaries and position griefwork outside of the "business" of the academy. The academy cannot bear loss; loss is unbearable in the academy. What ensued was an isolation and a longing to traverse the distance, to be home. Death had challenged the constitution of a professional identity predicated on a no-longer-acceptable level of displacement and loss.

Richard Johnson (1997) notes how, in his experience, "death put a desperate premium on remembering and at the same time seemed fatally to weaken this vital faculty" (p. 240). Away, without those who could remind me, without many things to remind me, my sense of self-in-relation felt deeply threatened. As Johnson also notes, "even personal memory depends on public support, the witness of others and social networks—at best a 'community' of rememberers, who are also, of course, partly constituted through their shared grief and memories" (p. 241). This longing for such a community of remembrance, inseparable from this place and the series of longings, responsibilities and obligations I felt toward it, led me home.

3. Nostalgia [Gk. *nostos*, to return home; *algos*, pain, longing]

As many migrants will attest, returning to a home place, when possible, can be no less fraught with anxiety than leaving. Returning is often a reexperience or an intensification of loss. Upon return I quickly learned that I was now "other," a CBFA (come-back-from-away), an alienating difference created in the juncture of leaving and returning, and noted through exclusions and silences—telltale signs of an estrangement mutually constituted and felt—a cruel twist to the insider-outsider dichotomy established in the original acronym, CFA. This odd "in-between-ness," feeling neither "one of them" nor "one of you," neither an insider nor an outsider—a common experience of returning migrants—ensures continued displacement. As such, it speaks to the intricacies and multiple forms of belonging that can emerge through migration. This in-between-ness also offers opportunities for insight, for seeing experiences through additional lenses and struggling differently with the contingent and conditional character of belonging. As Roger Bromley notes, "the in-between-ness of migrant identities, in the literal and metaphorical sense, both calls up, and calls into question, existing referential notions of cultural authenticity and traditional stable identity" (p. 67).

Geographies are sites onto which we may manifest our longings, but they cannot contain or fully assuage them. "Home" is a highly invested, politically complex mythology, a fiction fraught with contradictions. But it is a mythology by which we live, and in which we deeply invest. Reflecting on my own story has helped me understand some of its emotional dynamics and its cultural politics. Overwhelmed by compounded loss, I returned to Newfoundland to discover what I had lost. What I confronted was the profundity of loss this place had incurred in the time since I had left. The geography of loss and grief was dramatic, and nostalgia and melancholy were noticeable cultural habits (which I, too, was wearing).

The cod fishery moratorium of 1992—and the ecological disaster that was its impetus—is readily recognized as the major contemporary catalyst for the challenges Newfoundland and Labrador now faces. However, culturally, Newfoundlanders and Labradorians have been processing (or mourning) for decades (or, some would say, centuries) other "tribal losses" (Kessler, 2004, p. 150), many of which were mentioned in the introductory chapter: the loss of nationhood with confederation, the loss of so many community identities with resettlement, and the loss of trust with the sexual abuse crimes at the Mount Cashel Roman Catholic orphanage. Added to these more current losses is the weight of loss that reaches more deeply into our history and psyche: for example, the displacement of aboriginal peoples, the genocide of the Beothuck, and the accrued losses of those early White settlers. Culturally, it might be argued, we have not fully confronted nor productively processed these losses. Indeed, Ged Blackmore (2003), a researcher for the Royal Commission on Renewing and Strengthening Our Place in Confederation, calls some of these historically more recent developments part of "the new grief" (p. 329) of this place: "Once as a Dominion, then as a province, as communities, and, too often, as individuals we have never properly grieved what was lost. It is extremely important to understand that our disengagement from an effective grieving process helps account for our inability or refusal to confront our present social and economic conditions in a realistic, practical and humane manner" (Blackmore, 2003, pp. 327–28).

Like Blackmore, I am not arguing that these events provoked any singular or homogeneous experience or interpretation of those experiences, but, rather, that their cultural register suggests a deep, historically prolonged impact, which continues to resonate communally, in ways both collectively shared and contested. If "the management of change depends upon our ability to articulate the process of grieving" (Marris, 1986, p. 91), then we have been caught "between hope and despair" for some time. As a result, hope has been localized and sporadic, seemingly rationed against a broader spectrum of hardship and futility. As such, it has not seriously challenged or transformed old divisions and binaries of us versus them (forged through religion, locale, residence, race, citizenship, etc.) that are rooted deeply in our history.

Our overwhelming cultural responses to these events of loss echo those of people confronting loss and struggling to make sense and to reconstitute themselves in the face of difficult, life-altering circumstances. As both Johnson and Butler suggest, there is a challenge to identity presented in each of these historic moments of loss. Each has brought forward feelings of disorder, excess, anger, denial, intense debate, and "out-of-control" dispositions, feelings often born out of sorrow, guilt, and regret, and which resonate with what the psychoanalyst Melanie Klein referred to as "the depressive position" (Klein, 1976). From this depressive position, Klein argued, the grieving idealize the lost object (a person, a practice, a homeland, a way of life) as a way of both holding on to what has been lost and warding off the profound disorder loss threatens to bring to bear. This melancholy is both a means of working out (current and old) relationships and a form of repression that results in an inability to form new relationships and new attachments.

These emotive responses to loss—denial through idealization and subsequent ambivalence and melancholy—are also ways to relate to a collective identity.

As a moment of loss, migration can be charged with such dynamics. In Newfoundland and Labrador, the story of migration, particularly following the cod fishery moratorium, is widespread in its impact. Migration and the issues surrounding it are accompanied more often by silence, anger, resentment, or resignation than by any real concerted action. Despite fifteen years of unrelenting outward migration, with the exception of the 2008 Youth Retention and Attraction Strategy that has only recently begun its exploratory research, there is little concerted government policy that socially, culturally, or educationally addresses the phenomenon, its causes, and its consequences—this despite the challenges to community institutions created by its social consequences. Irish scholars (Mac Einri, 2001) document a similar silence in relation to Irish migration, in particular, that of rural Irish youth. In her comments on the impact of Irish emigration, Mac Einri (2001) sheds some light on our situation in Newfoundland and Labrador:

> With a strong family-centred ideology on one hand, [the Irish] had to cope, on the other, with the reality that we were rearing children for export....We dealt with [this classic double bind] largely by denial, accompanied by a number of other coping strategies, such as a certain resentment of those who had emigrated (and even more of those who returned). In general, Irish responses are probably best characterised as an attitude of fatalism and an ideology of victim-hood, coupled with total official inaction over many decades. A sentimental attitude toward family, exile and the "world wide Irish family" enabled us to ignore the realities. We took refuge in an idealising vision of ourselves, rooted in a place which was by definition unchanging—the past. Above all, we denied the class-based reality of involuntary migration—the choices open to the disadvantaged were and are simply more limited, or nonexistent, compared to those open to the rich. (p. 4)

The extent to which these insights resonate within the context of Newfoundland and Labrador is striking: the family-centered ideology, the heartbreak of raising children to leave, all-too-common attitudes of fatalism and victimization, cultural idealization, and melancholia as a habit of denial.

Grief is cultural, but cultures, like individuals, must also grieve. Cultural depression and melancholy—and its symptoms and effects—are manifestations of damaged and disrupted cultures that have been unable to process historical trauma and loss or to reconstitute themselves in an empowered way in the present, for the future. Note the seemingly endless debates about fisheries stocks, quotas, and management. Listen to high school students express in taken-for-granted ways the disadvantage of their province and the inevitability of their leaving following school, a structurally embedded educational effect that Michael Corbett (2007) calls "learning to leave." Watch newly minted teachers, nurses, and skilled trades persons disappear elsewhere. See the effects of family fragmentation on the faces of those who take and meet Air Canada's FortMac Direct, an "airbus" for the "three and one" and "six and two" shifts—those migrants who work in Alberta for three or six weeks but then "live" in Newfoundland and Labrador for one or two, economic migrants who both leave *and* stay at home. Hear municipal leaders speak of the erosion of community infrastructure and morale. Observe the unending struggles of our aboriginal peoples for recognition and autonomy. See the onslaught of museum mania as the new rurality. Note the verbosity, cut with blame, victimhood, and resentment, that clogs the airwaves of radio talk shows and fills pages of letters to newspaper editors. Count the closures—of schools, fish plants, churches, and small businesses. And, then, consider how—*despite these losses*—the prohibition on grieving intensifies nostalgia as the only permissible form of mourning while also reproducing well-circulated myths of this province and its people.

Despite having been away from Newfoundland and Labrador for two decades, I did not see myself represented in the outward migration figures announced with each new census, in part, because, like so many other Newfoundlanders and Labradorians who move away, I had not spiritually left the province. In some odd way, despite being "away," I felt myself still "here." As Nancy Huston (2002) notes, "every expatriate has the conviction—deeply rooted in her subconscious and regularly rejected as preposterous by her intellect—that a part of herself, or, rather, *another self*, has never stopped living *back there*" (p. 91). While away, I did not stop longing to return; my return felt inevitable, even if its timing was uncertain, even though, now, return seems temporary, as much a pending departure as a longed-for arrival. As I have written elsewhere (Kelly, 1993; 1997), my desires around Newfoundland were contradictory, marked by ambivalence: I desired to be here where I was not entirely at home and to be away where I could be more at home with myself. Considering the impact of loss, personally and culturally, has helped me to recognize the cultural constitution of my own ambivalence and the deep and far-reaching implications of its basis in loss, and, in particular, in melancholy as a response to loss. These considerations reshape a personal relationship to culture and, as well, to profession. Together, they propel a reconstitution of my educational work as it relates to this place.

EDUCATION [L. *EDUCARE* TO LEAD FORTH]: WHAT AM I DOING HERE?

Other regions give us back what our culture has excluded from its discourse.

M. De Certeau, *The Practice of Everyday Life*

What has this excavation of loss and memory mobilized? What are the implications of confronting the disjuncture between longing and materiality, between what one desires and what one has? Between where one was and where one is? Between where one is and how one belongs? Part of my longing to return to Newfoundland—and part of my reluctance to leave in the first place—was a deep desire to contribute my labor, my life work, to the place. While away, I experienced this longing as loss. One of the ways in which we contribute to communities is in our participation in them through the work we do, the contributions of our labors to community sustenance and growth. A sense of place and belonging is deepened through such efforts. For me, the work I do—academic work—must have some tangible vision of the communities in and for which it is done; education—the specific academic area in which I do most of my work—bears an especially grave responsibility to community and the hopes, dreams, and struggles that form it. When communities are struggling with loss and change, education is crucial to the processes by which these struggles can be addressed. Despite the challenges of the contemporary world, educators continue to teach to a pseudonotion of permanence—whether of knowledge, history, culture or humanity. What I believe educators must begin to address in a more concerted way is the nature—and hopeful inclinations—of life in the context of change, uncertainty, and loss.

An established aversion to grief and a history of unarticulated and unresolved grief are intensified through educational silence (Behar, 1996; Kessler, 2004). Indeed, such silence implicates education in the psychosocial difficulties of cultural loss and change. There is a relationship between the problematic separation of emotion and intellect, which makes death "other" and obscene in academic institutions, and the reluctance of education to address more fully and meaningfully issues of loss, change, and cultural renewal. It seems important to find new ways to engage issues of "life-and-death" importance to communities, to address the pain, grief, humiliation, struggle, *and* hope about which we must begin to speak differently. Accepting or attempting to maintain the status quo is not a productive response to cultural crisis in that it is a form of refusal to acknowledge what is lost. Crisis, by its very nature, challenges the status quo. The desire for renewal and revival—what Jonathan Lear (2006) calls "radical hope"—must transcend resistance. From this different and deeply hopeful place, academics and educators might resist the urge to *dissociate* (through silence, exclusion, isolationism, and intellectualism) from the cultural losses and struggles that surround us and in which we are implicated and against the binaries and exclusions that help keep us (comfortably but unconscionably) separate from the changes which abound—that is, comfortably numb.

Not unlike many other decimated places, our most noticeable province-wide response to cultural loss, thus far, has been to develop a more vibrant tourism. Reinventing rurality as a tourist commodity is a band-aid on the gaping wounds of history.

More dangerously, through their idealizing tendencies, many tourism practices appear to foster memory but, instead, can encourage a kind of forgetting, often rooted in melancholy, that supports nostalgia and fuels distortions and repressions. A more radical remembering is required. For this to happen, a country, a culture, must first of all acknowledge its losses and disavowals, ask what they mean, inquire into the nature of its collective responses to them, and recompose itself collectively, across difference. William Pinar (1991), in writing about the American South, captures the cycle of unresolved loss. He argues that "pain repressed produces pessimism, which in turn supports nostalgia, which in turn supports provincialism, conservatism" (p. 176). There is an impatience brewing, slowly and steadily, toward this stunted mourning. Ray Guy captured a particular cultural ambivalence in his recent description of Newfoundland and Labrador as "half heaven and half hell" (Guy, 2007, p. 5), the heaven being the best inclinations of its people and the "hell" being their worst, those borne of some of the regressive and counterproductive tendencies of the place, those steeped in nostalgia and melancholy and that are, as a result, prohibitive of new attachments and change. Likewise, Sean Panting (2007), an artist living in St. John's, the provincial capital, comments on the implications of a nationalism that emerges through dysfunctions similar to those to which Pinar refers: "It can make us resistant to change, resentful of outsiders and, occasionally, downright xenophobic" (p. 19). Or, said differently, unaddressed grief leads to blame, depression, despair, dysfunction, and, as their result, various forms of aggression toward self and others (Kessler, 2004, p. 138; Blackmore, 2003, p. 328). As Wendy Brown (2006) reminds us, "aggression is what emerges in the spaces of unmourned losses" (p. 31). The choices of targets of our aggression can betray the nature of our historic and contemporary exclusions. Yet, a wild impatience with the manifestations of melancholy will not mobilize more effective forms of grieving. Nostalgia and melancholia must be understood not as fixations but as symptoms of unmourned or disavowed losses.

In the context of such unmourned cultural losses, categories of exclusion threaten to emerge more vividly as the Province becomes home to many more immigrants. Claims that Newfoundland and

Labrador is a homogeneous society are too commonly flaunted and are deeply flawed. Nonetheless, this myth of sameness functions effectively to refuse, deny, and suppress difference. Challenges to exclusionary constructions of difference—in particular, racialized difference—are not common public or educational currency, nor is the need for them widely acknowledged. Some of this may be attributed, in part, to the fact that outward migration is, at this historical juncture, a greater social force than immigration, the latter still for all slowly but steadily challenging the largely White social constitution of Newfoundland and Labrador identity. The work of Mac Einri (2001) on Irish identity is, again, pertinent here: "Emigration translates in social terms as *absence*, rather than *presence*. To a large extent this has meant that certain questions most relevant to receiving societies—the whole field of community and national identity formation—[are] seen in a radically different light, or perhaps avoided entirely. To put it another way, society, without substantial immigration, could continue to be constructed as monolithic" (Mac Einri, 2001, p. 4).

One of the consequences of such massive outward migration is the need to encourage immigration as a means to sustain social institutions. The "brain and brawn" drains have taken their toll on communities who clamor to find professionals, and skilled tradespeople and workers, to meet their own needs. In March 2007, the government of Newfoundland and Labrador announced the formation of an Office of Immigration and Multiculturalism, which has a broad mandate to attract and retain new immigrants to the province. The public response to this announcement is fraught with the anxieties, complexities, and exclusions of a colonial culture. Despite numerous aboriginal peoples and many "diasporas within," the construction of cultural identity, as a colonial identity (in empire) and as a "regional" identity (in nation), has refused this heterogeneity. As a result, the aboriginal peoples of the province, for example, have been subjected to unconscionable suffering. But aboriginal peoples are now renewing their communities as proud inheritors of a rightful place—too long denied—in the Provincial makeup, largely through acknowledging and working to heal from such devastating loss. Changing times for this province will require that its reputation as caring and welcoming be accented by an education that takes the lead provided by aboriginal peoples, one that probes the basis of our historical losses and disavowals so that this caring reputation can be realized in a deep and meaningful way. Artists, writers, academics, and educators all have a role to play in such cultural renewal.

Conclusion

Migration, with its foundation in loss, dramatically challenges notions of identity and community. The migrant story, from the most seemingly mundane and ordinary, as partially told here in this essay, to the more elaborated accounts available across cultures, provides the place from which to study the structure of mourning that frames migrant stories and to gain insight into the human condition in a new century. In the migrant story are expressed some of our greatest hopes. However, the migrant story also points to a culmination of losses, many of which are accrued through the naturalizing of deeply problematic social priorities. In this twenty-first century, as the nomadic citizen replaces the permanent inhabitant as the dominant cultural figure, there is reason to pause, to assess what slips away in such movement and change, and, as well, to reconsider what is acquired through new, renewed, and contingent attachments. One of our deepest human impulses is to protect ourselves from loss, despite its mark on all our encounters. Comfort, connection, and insight in the face of loss are essential to prevent counterproductive and even destructive responses and to promote healing, growth, and change. These insights are essential components of the consciousness of contemporary cultural educators in this, "a new sort of world in which diaspora is the order of things and settled ways of life are increasingly hard to find" (Appadurai, 2003, p. 14). Within education, the migrant story must be heard, analyzed and understood, for migration, that "quintessential experience of our time" (Berger, 1972, p. 55), indelibly marks those we teach and those who teach. Acknowledging the mark of this experience enables us to proceed with the kind of concerted care that supports informed hope, inspires more thoughtful encounters of difference, and encourages meaningful change.

POSTSCRIPTUM: TRANSMIGRATION[3]

As he left the church, filled to capacity on this midweek morning, and walked past the lingering crowd, he thought of how many of these memorial services he had attended in the past years. It seemed to be the new way of doing right by the dead: two funerals—one away for those *there* and one home for us *here*—the body already ashes by the time it fulfilled the rite of return. It has to be this way. Home holds your presence long after you have gone. Places never really let you go any more than you ever really let them go. There is still remembrance here, some vestige of hope, he supposed. Not that things would ever be the same. Too many gone; too much passed. Truth be told, it has always been the way that dreams could take you away—if you managed to let yourself leave, or if they did. And being in the wrong place at the wrong time could happen anywhere—even here. Bad timing is just easier to take when you do not have to relate to it long distance. If following your dreams kills you, you want someone there, in the end, who remembers where the dreams were born, even if it was in a place that could not hold them—or you. You want someone to remember what it took for you to chase your dreams—and all those odds you beat down. Never mind the ones that beat you down. You want someone to remember not just what those dreams made of you, but what making them took from you, too. And you want someone who will make sure you get back home. In the end, nothing takes home from you.

Chapter 2

Writing "The Distance Home"
Migration, Mourning, and Difference

Introduction

And who would say, to watch him sleeping there, one hand trapped under his face, his dimple squashed, that this little boy will grow to be a man, like the rest of us, whose surest legacy is loss.

Lawrence O'Toole, *Heart's Longing*

Early in his memoir *Heart's Longing: Newfoundland, New York and the Distance Home* the author Lawrence O'Toole establishes the tenor of everyday life in the imagined community of the title, Heart's Longing, a close resemblance to his birthplace of Renews, Newfoundland, a small fishing community on the southern Avalon Peninsula of Newfoundland and Labrador. In these initial pages, O'Toole focuses on one day as he provides glimpses of the comings and goings of various people in the community. But this is no ordinary day in Heart's Longing, for a death has struck one of the families, shifting the emotional landscape of the community and drawing all eyes toward the sisters, Dottie and Alverne Reddy, whose sister has died suddenly the day before. O'Toole describes the sisters in their home the morning following their sister's death, noting how, as they go about preparations for the wake and funeral, "neither survivor can find her grief, as if she'd somehow misplaced it" and how "each privately thinks how queer it is she cannot find her grief for her dead sister" (p. 11). In many ways, O'Toole's memoir can be read as a quest to confront a *double entendre* of the "queerness"[1] of grief—its elusive, haunting, and often idiosyncratic character, which can be accentuated by sexuality and difference—the long and difficult mourning that is finding and negotiating grief, the "distance home" of the title. This distance is physical, emotional, and spiritual, and intensely confounded by displacement and difference. The narrative is a pilgrimage of loss and reclamation and a testament to grief's haunting—how it is that, when we seem unable to find our grief, it invariably finds us and makes a home within. In these ways, the visit of death to the community of Heart's Longing in these opening pages is a fitting mise-en-scène for this memoir.

Cultural texts are opportunities to ask certain questions and to inquire into particular politics and affects. O'Toole's memoir has personal meaning, given how his experiences resonated with those of mine outlined in the preceding chapter. Despite our difference of gender, we have many commonalties: the struggle to reconcile one's sexuality; the deep love of and alienation from place; and the recognition of mourning as a central part of coming to terms with both these difficulties. In addition to these personal interests, I would argue that two characteristics of this memoir help make it a potentially rich object of study in a broader context of an educational discourse of loss and place: first, the memoir captures well some of the complexities and contradictions of leaving, longing, and belonging; and, second, it exemplifies an explicit negotiation of mourning, melancholia, and difference. In these senses, I argue that, as a multilayered, artistically evocative study of migration and the vicissitudes of leaving and return, and of reconciliation and renewal in the face of loss, the memoir is what Paula Salvio calls "a narrative of reparation" (Salvio, 2006, p. 84) through which complex, contradictory, and ambivalent relationships to place, to self, and to others are explored. The memoir is also a study in writing itself as reparative. Using O'Toole's memoir as an analytic focus, then, I pursue the following questions: What does this memoir suggest about reconciliation of the losses that attend migration? What structure of mourning does this story incorporate? What does the memoir suggest about place attachment and identity and place? What does it suggest about the experience of emigration and difference? And, how might these insights inform thinking about cultural loss, change, and renewal?

Heart's Longing

How far I've come, and yet how little a distance it is.

Lawrence O'Toole, *Heart's Longing*

Heart's Longing was published, to little acclaim, in 1994, just two years after the historically unprecedented and economically and culturally devastating moratorium on the five-centuries-old northern cod fishery in Newfoundland and Labrador, its historical *raison d'être*. In the period following the moratorium and subsequent closure of the fishery in 2003 because of catastrophic stock depletion, there has been massive outward migration from and depopulation of hundreds of former fishing communities, similar to Heart's Longing, which pepper the coastlines of the province. This same time period was also one of devastation by AIDS-related deaths in North American gay male communities. It is against this backdrop of O'Toole's two very different and beloved, yet both gravely imperiled, island communities and their losses that his memoir is emotionally positioned. O'Toole, a widely published writer,[2] recounts, in this memoir, how, as a young man, he left the fishing community of Renews, Newfoundland and Labrador, and what he calls "the contemptible familiarity of its confines" (p. 5) to find himself at home in another village on another island, Manhattan. He writes of the experiences that led to what he calls his "reverse pilgrimage" and how "I reacquainted myself timorously with who I'd been to find out who I was" (p. 33). In so doing, he learns that the "familiarity, which used to breed contempt, is now that which I seek" (p. 113) in order to find home, a place of belonging—the mythically conceived but emotionally real Heart's Longing of the title.

That Heart's Longing is a creative amalgam of various realities underscores the rub of nostalgia: Heart's Longing is an idealized, romanticized version of the real, and any real, actual physical return is impossible. Its creation, then, is a narrative choice that resonates with the emotional disjuncture between memory and reality that undercuts any nostalgized version of home. As such, it is an important "third space" onto which redemptive impulses can be projected and in which alternative longings can be examined and the meanings of and longings for home explored. The mythological place of Heart's Longing is populated by characters who are also amalgams—fictive versions of socially and psychically real selves, which include the multiple sides of the narrator, as well as a sampling of the real voices and dispositions of a place. There is Jipp Parnell, who has left the Roman Catholic priesthood after nineteen years and has now returned to Heart's Longing, his home and "the only place from which to start all over" (p. 88). O'Toole writes that "Jipp Parnell is the character in Heart's Longing with whom we will have the most dealings. His story, as its core, is very like that of any Newfoundlander who has left and seeks to return" (p. 78). There is young Derm O'Dairn, "a typical Newfoundland guide from back then" (p. 47), a character based on the narrator's brother, who is a generous, enigmatic guide to Heart's Longing. There is Shadow, the homeless dog, whom Derm can be heard calling or admonishing, meaningfully, to "Go home" throughout the story and whose name suggests the Celtic inclination for melancholy, "to look toward the shadows rather than the sunlight" (p. 197). Finally, there is the additional supporting cast, each of whom has an indispensable, even if sometimes small, place in O'Toole's story.

O'Toole frames his memoir not around migration but willful travel—movement as a search for meaning propelled neither by state and national politics nor economics—which, while a more autonomous and, perhaps, exotic metaphor, is one that does not exempt him from experiences of displacement and exile, borne as much of his sexual identity as of a profound relationship to home and place. His creative amalgams are the means by which the complex emotional grammar of migration, loss, and mourning is structured and the central themes of the memoir are realized. They are the narrative means by which are shown the author's response to—and navigations through—the difficulties of memory and of presenting a past "hopelessly distorted by nostalgia, denial, forgetfulness, shame, and some of the other little mercies of memory" (p. 6). Loss informs the story as both personal experience—the real losses accrued through migration and change—and cultural bequeathal—the affective history of people and places—and their deep and profound intersections. In this sense, the memoir is a study of identity and place and their emotional vicissitudes and legacies.

Emigration and Loss

[T]o emigrate is always to dismantle the centre of the world, and so to move into a lost, disoriented one of fragments.

J. Berger, *Keeping a Rendezvous*

Real dislocation, the loss of all familiar external and internal parameters, is not glamorous or cool. It is an upheaval of the deep material of the self.

E. Hoffman, "Wanderers by Choice"

In their discussion of the persistent melancholic character of Asian American assimilation struggles, David Eng and Shinhee Han (2003) begin from the assumption that the experience of immigration is based on a structure of mourning (p. 352). Their particular focus is racialized immigrant groups for whom assimilation into White society is fraught with contradictions and is always ever partial and incomplete. The partially assimilated cultural subject is one who is haunted by previous identities, places, and forms of belonging that have been displaced through immigration. Rereading Freud's well-known distinction between mourning and melancholia (Freud, 1917/1989), reviewed in the introductory chapter, Eng and Han explore the incomplete and unresolved mourning that is melancholia as "a depathologized structure of feeling" (p. 344) and a not-uncommon feature of many collective or cultural, as well as individual, formations. In this sense, along with many other contemporary scholars (Butler, 1997; Cheng, 2001; Gilroy, 2005), they counter the prevalent pathologized notion of melancholia and, instead, enlist it as an affective feature potentially ripe with political possibilities.

According to Eng and Han, the impossibility of full assimilation can lead to a melancholic subject, for whom such haunting by and longing for the lost objects of love (from the home culture) inhibits full investment in new objects. This melancholic character, these unresolved losses, they argue, are bequeathed; they become intergenerational, with younger generations becoming "place holders" for the unresolved losses of their ancestors. Eng and Han argue that this displacement of loss has political potential, for it also signals a "displaced reside of hope for the reparation of melancholia" (p. 354). This "ethical hold" (p. 365) on treasured but lost, forsaken, or threatened identities is framed by a complex desire to maintain or to rehabilitate a relationship undervalued, injured, or abandoned—a relationship of identity and culture. In this sense, reparation is both a creative reanimation of what has been lost, an expression and extension of bequeathal and legacy, and a fierce and binding display of loyalty and love.

Although constituted differently, politically, the migrant offers another point of entry to this complex positioning, as O'Toole's memoir demonstrates. The migrant, too, is caught between displacement and a configuration of home place constituted by history, experience, myth, and memory, which together can fuel a longing to return. But return more often offers a reminder of the new differences that exist, necessarily, outside the register of nostalgia. For example, on a visit home, O'Toole remarks, "While recognizing my renewed kinship, I nevertheless still felt like a visitor who had come back to look and then leave. My speech was different. And so were my clothes...My profession—a journalist—was pitched beyond the understanding or interest (mostly interest) of the majority of people in Renews. Paradoxically, to have done well outside the perimeters of Renews means to be 'grand' and not really a part of it. I was a stranger at home" (pp. 36–37).

The loss is palpable: "There was a time when I knew everybody and vice versa; and I was greatly uncomfortable with that fact. Now I'm uncomfortable that I'm a stranger to many, and they to me" (p. 41). Nor does home remain the same. As O'Toole quickly adds, "Home had become strange, too," now more modern, yes, but smaller, too, and quieter, its vibrancy cut away by depopulation and decay. Nancy Huston (2002) captures this split the expatriate feels: "You've got one life *here* and another *there*....*Here*, you set aside what you used to be.... *There*, you set aside what you've become" (pp. 10–11). Through return, loss is not recouped, but doubled, its intensification an entry point to the melancholic subject.

Where might such losses (continue to) resonate? What—or, more importantly, who—are the placeholders of such unresolved losses? In his penultimate chapter, O'Toole says that Heart's Longing is "a place that existed a long time ago, a place to tell your children about. But since I will have none of my own, I am telling other children" (p. 192). He then dedicates his book, a dedication to a time, a "for when," not a "for whom": "This book is for when there are too few people left to tell stories of the fairies and the wren and the fools and all those things that never happened quite the same way in any other part of the world through which we make our way every dumbfounding day we are given" (p. 192).

"When" portends a time of ultimate loss, for which story remains its final trace, a time home is no longer conceivable, except through story, and when another place, another home must be secured—when the story and the storyteller are irrevocably distanced. Such severance does not preclude an affective legacy. Rather, it raises questions about the nature of such legacy and the place of story in the reconstitution of place and identity.

IDENTITY REELS[3]

Identity most often rhymes with marginality.

E. Probyn[4]

I am what I am through my own efforts, surely, but I am who I am because of where I come from.

Lawrence O'Toole, *Heart's Longing*

As well as a reckoning with loss, O'Toole's memoir may be read as a reckoning with identity, identity reconstituted through migration and the struggles of difference. As such, it calls into question the psychic and social bases of identity, its claims, affirmations, disavowals, and consequences. Stuart Hall's work on identity offers an important frame to explore identity as it relates, in this case, to Newfoundland and Labrador, in particular, to O'Toole's memoir of place.

Hall distinguishes between two conceptualizations of cultural identity: identity as essence, and identity as position. The first, an essentialized notion of identity, is identity framed through a shared history and ancestry out of which is formed a notion of essential or true self. It is solidified through cultural practices that reiterate this cultural essence as part of the conditions of community and belonging, despite or in denial of what Hall calls "the shifting divisions and vicissitudes of our actual history" (p. 234). Framed within largely acritical claims to the value of heritage, such identities are culturally stabilized and idealized, their strength a potent political tool in social and cultural struggles.

In the context of Newfoundland and Labrador, it is this first conception of *identity as essence* that informs cultural discussions of "a true Newfoundlander" and predominates in media and, even, school textbooks such as *The Newfoundland Character*, a collection of writings used in English Language Arts curriculum in the province in the 1980s. The opening lines of this textbook capture its purported purpose, "to develop the concept of the Newfoundland character" (Ryan & Rossiter, 1984, p. 1), a purpose reiterated in a note to students that the text will "give you an idea of the kind of people Newfoundlanders are by looking at their past history, culture, and their struggle with the sea and the land" (Ryan & Rossiter, 1984, p. 3). This "Newfoundland character" is a character of adversity bred in a culture of hardship and tragedy: "The Newfoundland character is not a person as such. It is the sum total of the qualities that come through in the way Newfoundlanders have met the difficulties and the challenges of living on an island whose climate at times is harsh and uncompromising, whose history for centuries was not favourable to settlement, whose merchantile barter system kept money from going into the pockets of fisherfolk, and whose isolation kept settlers together in closely knit communities while at the same time keeping them apart from other communities" (Ryan & Rossiter, 1984, p. 1).

This mythical character, cast from the circumstances of history, geography, and economy, maintains its powerful hold on the present, mobilized and reiterated in response to contemporary struggles and political interests. For example, in a 2007 press release from the provincial Progressive Conservative Party led by current Premier Danny Williams to launch a provincial election, the choice of campaign slogan, "*Proud. Strong. Determined.* And *the Future is Ours*," was explained:

> The Campaign slogan for the PC team embodies the essence of the people of the province. "We are *Proud. Strong. Determined.* And *the Future is Ours*," said Premier Williams. "These powerful words are not just about our team of candidates in this election; they are reflective of the people of this province as a whole. Nowhere on earth will you find such fierce pride, such strength of character and such determined spirit, than right here in Newfoundland and Labrador. With these characteristics as our foundation, the future is indeed ours for the taking. *Proud. Strong. Determined. The Future is Ours* encapsulates the most vibrant and self-sustaining characteristics of our people and this place we love so much."

On another occasion in the same 2007 election campaign, Williams created much debate when he reiterated an oft-heard reference to Newfoundlanders and Labradorians as a "race" of people.[5] While such references have a complex history,[6] the mobilization of such exclusionary language to feed nationalist fervor for political gain, while not uncommon, is both problematic and dangerous.

Indeed, against the backdrop of massive outward migration and, as a result, much-needed immigration, other initiatives by the same government to instate the province's first Office of Immigration and Multiculturalism appear to counter these earlier claims. The office acknowledges in its *Policy on Multiculturalism* (Government of Newfoundland and Labrador, 2008) that "culture, as a way of life, is dynamic, learned and ever changing" and, as an extension, so, too, are the cultural identities constituted in relation to and as part of it. It realizes, however, that if its principles of respect, equality, collaboration, and inclusive citizenship are to be realized, it will have to do educational work.[7] Such work would have to counter the enshrined notions of citizenship already embedded in the cultural imagination, available to be tapped and maintained by political interests that would benefit from their utilization.

In identity politics, such mythic essence and cultural conservatism complement in a potent cocktail, served with highly select offerings of history, tradition, and culture.[8] Such essentialized notions of identity wield power because, as Ing Ang (2000) points out, "they *feel* natural and essential" (p. 2). In relation to Quebecois, another widely debated notion of identity in Canada, in a province—Quebec—in which identity politics are often at the forefront of public discourse,[9] Elspeth Probyn (1996) notes, "Listening to everyday expressions of Quebecois identity, I am reminded of its theoretical impossibility; I am reminded that people do routinely accept it as an actuality. Thus, while it may be my experience that having *an* identity is an impossible idea, it is something that nevertheless circulates as a feasible goal and increasingly as evident fact" (p. 71).

Hall's second conception of *identity as position* acknowledges, theoretically, that who we are is an amalgam of points of similarity and, also, of deep and dividing differences that debunk any coherent notion of cultural identity on which essentialized notions rely. Hall speaks of a dialogic relationship between "two axes or vectors, simultaneously operative: the vector of similarity and continuity; and the vector of difference" (p. 237)—the unstable, shifting and often conflicting intersections of continuity and discontinuity and rupture. This relationship, Hall argues, is "not an essence but a *positioning*" (p. 237) in relation to what is inside and out. As a gay man writing about Newfoundland as his place, too, O'Toole challenges the abjection that would position him outside of "the character" of this place. For that reason, O'Toole's memoir provides insight into the intersections of these vectors through both a claim to a place within a particular history and a place as a gay man emerging from that history.

Ing Ang draws on Hall's work on identity to analyze the emerging context of identity struggles in Australia; she describes identity as "a confrontation of the past and the future, a tug of war between 'identity' as essential being, locked in (an image of) the past, and 'identity' as open-ended becoming, invested in a future that remains to be struggled over" (Ang, 2000, p. 9). Similar tugs-of-war are evident in Newfoundland and Labrador in the public domains of newspaper articles, open-line radio talk shows, and Weblogs, as well as in more intimate exchanges at grocery supermarkets, in pubs and restaurants, and around kitchen tables in smaller and larger communities. Increasingly, identity feels linked to destiny in new and ever more complex ways. Caught in a veritable sea change, such rhetoric attempts to stem the tide and build a breakwater against loss and change. Despite the suggestion of her essay title, "Identity Blues," Ang does not explicitly explore the melancholic features of this identity struggle and the forms of loss and denial out of which it arises. Yet, with its associated aggression, it would seem that melancholia feeds its intensity.

It is left to Judith Butler (1997) to remind us that identity based on repudiation produces a particular form of melancholia that requires an expressive language in order to move on. The petrification or immobility that is a hallmark of melancholia is vulnerable to the strictures inherent in claims to an essential identity. An essentialized identity is a disavowal of difference, a zealous, excessive overidentification. As Butler (1997) points out, disavowal in the form of overdetermined identification "raises the political question of the cost of articulating a coherent identity position by producing, excluding, and repudiating a domain of abjected specters that threaten the arbitrarily closed domain of subject positions" (p. 149). This question is central to a project of well-considered change that attunes to all forms of disavowal and, specifically, to forms of disavowal in relation to gender and sexuality—that is, of both a heterosexual and a gay melancholia—and the prohibitions and attachments of which they are socially and psychically constituted.

Queer Departures

> *When certain kinds of losses are compelled by a set of culturally prevalent prohibitions, we might expect a culturally prevalent form of melancholia.*
>
> Judith Butler, *The Psychic Life of Power*

The opening line of O'Toole's memoir consists of four words: "I want to belong." The closing line is but three: "I am home" (p. 200). The story that traverses the distance between these two sentences, or "the distance home" (p. 1), is a struggle to reconcile the meaning of belonging, to find solace and joy in who one is and what one makes of it—to be at home with oneself. For O'Toole, this struggle to reconcile his "endless night" of being gay, an alcoholic, and a Newfoundlander—all liabilities of identity within the dominant frames of North American culture—is marked by a tension between being and becoming and leaving and returning. Elspeth Probyn (1995) captures the ephemeral quality of belonging when she states that belonging "carries the scent of departure—it marks the interstices of being and going" (p. 2). The reconciliation to which O'Toole gestures, although no less fleeting, captures the hard work of constituting these "queer belongings" to which Probyn refers (p. 15) in the context of widespread cultural disavowal.

How is the migrant experience complicated by such queer departures, such queer quests for belonging? Gayatri Gopinath (2003) argues that "nation and diaspora are refigured within a queer diasporic imaginary" and that "nostalgia as deployed by queer subjects is a means for imagining oneself within spaces from which one is perpetually excluded or denied existence" (p. 275). Probyn, writing about her own queer regret and nostalgia—common themes in O'Toole's memoir, as well—suggests similarly: "Regret and nostalgia form part of the movement of belonging; they regulate the speed with which images pass by. They slow down certain images so that I can grasp a fleeting sense of belonging" (1995, p. 15). This ephemeral nature of belonging enhances its allure and fuels further the desire to belong. Yet, in the quest for queer belonging, it can also mobilize a repudiation of "other" desires in order to feign cultural allegiance.

To repudiate desire in order to feign belonging creates not necessarily a false but certainly an incomplete sense of home, a home built on disavowal and loss. And neither does moving with the current of queer desire recuperate loss if it enacts its own repudiations, thereby creating a greater distance from home. When O'Toole writes about his move into a public world of queer identity, he marks it as a departure from home. "I knew that the moment I entered that [gay bar in Toronto] I would be divorcing myself even further from where it is I come from and that I would be, as the spiritual says, 'a motherless child...a long, long way from home'" (p. 109). Yet, in describing the relationship of home and difference, O'Toole is generous:

> In Heart's Longing, a code of respect operates whereby no one takes advantage of someone else's hardship, cross, flaw, failing, limitation or "difference." This is not to say that Newfoundlanders are a more innately tolerant people. They are not. But a code of behavior has been cultivated over the years whereby limitations of one sort or another demand sympathy rather than judgment....It is not uncommon to hear it said of them, "God bless the mark." As in ancient tribal life, a difference or mark is a sign of benediction, however oddly manifested, from the gods. This is the village mentality at its kindest, which is to say its best. (p. 85)

And there are many in Heart's Longing who bear the mark. O'Toole tells of village eccentrics, gay priests and community members, and persons who are mentally challenged. Among these, Katie Ann stands out as the symbol of the misunderstandings created when fear of difference guides the meanings we make of one another and, as well, of all that is lost when the bonds of our common humanity are thwarted, distorted, and diminished. Isolated in the village by her difference, accentuated by her appearance and her reclusiveness, and that of her brother, with whom she lives, Katie Ann is a woman of as much generosity as mystery. She teaches the young Derm, one of O'Toole's alter egos, one of his earliest lessons in love and acceptance and, in surprising ways, she allows him small moments of joy. Through her wistful, quiet devotion to an undisclosed, deceased love, Katie Ann also teaches Derm something about memory, its melancholy, and our responsibilities toward it. But her character is also a reminder of the limits of melancholia, its traps and distortions, and the rigid parameters it can place around a life.

Yet, this benevolence of a community toward difference is as inconsistent as it is idealized. It is also coercive and disavowing, qualities that O'Toole implicitly captures in his account of his home community and its attitude toward queer identity. He notes, "In Heart's Longing homosexuality enjoys a surprising amount of tolerance" and that the word *queer*, culturally, is not "a term of opprobrium" and "has nothing to do with sexuality" (p. 84). Later, he says, "As narrow-minded as Newfoundlanders can be about certain other things, they are surprisingly accepting about gayness. When I was growing up in Renews, the people who were gay, albeit extremely closeted, were not shunned or ostracized. Everyone *knew*. God bless the mark, as my mother would say, and I'm sure God does" (p. 133).

The silence as disavowal—"the love that dare not speak its name"[10]—is a prohibition of voice and being. The coercion that is silence as benevolence solicits complicity to purchase a partial and fleeting sense of belonging. O'Toole reiterates this knowing as disavowal when he writes, "I *knew* I was gay from as early as I can remember" (p. 110). His memoir is testament to a "queer" melancholy, and the consequences of melancholic positioning. The memoir avows "the number who lived faithfully a hidden life"[11] and shores up the place of those cultural subjects whose psychic and social insistence on and adherence to sanctioned prohibitions are implicated in their hiding.

While acknowledging another loss, O'Toole links this denial to a series of disavowals that must be mourned. Describing an unborn, lost son, conceived in an early heterosexual relationship, for whom he had not properly grieved, O'Toole writes, "I have never properly grieved for him, my lost son who almost once was. But maybe I have [grieved] without knowing it. While grieving the loss of something else in my life—childhood, a lover, a friend—I probably cried for him, my lost son....Surely, too, I must have grieved for jettisoning my roots all those years, and the depth of that feeling, for most of my life, must have been fathomless. I was grieving for *me*, the man of my dreams, the man in the mirror" (O'Toole, p. 110).

The roots to which he refers, while ambiguous, seem to include here relationship to place, as well as his relationship to a nascent sexual identity. Here O'Toole calls up Klein's assertion that "early mourning is revived whenever grief is experienced later in life" (Mitchell, 1986, p. 147) and that these later anxieties of loss precipitate early infantile processes of reconciling loss through rebuilding a loving inner world—what Klein calls the reinstatement of lost objects (pp. 166–67). It would seem that the memoir, as a reparative gesture of this Kleinian sort, functions to rethread these connections—to people, place, and self—and to now enable successful mourning.

Writing and/as Reparation

As noted in the introductory chapter, in her work, Melanie Klein links loss, mourning, reparation, and creativity. She argues that a reparative urge is borne of a confrontation with loss of a loved object, ideal, or person, and it "includes the variety of processes by which the ego undoes harm done in phantasy, restores, preserves and revives [lost] objects" (Klein, 1986, p. 48). According to Klein, this reparative character of mourning, dominated as it is by love, can be quite productive, prompting creative expressions through which the lost object is lovingly reinstated, and through such successful mourning inner harmony can be restored. Klein further adds that "any pain caused by unhappy experiences, whatever their nature, has something in common with mourning…and encountering and overcoming adversity of any kind entails mental work similar to mourning" (Klein, 1986, p. 164). As many scholars have noted, such reparation resonates individually and culturally.

The relationship of writing and reparation to survival and reconciliation is echoed by many writers. Alice Munro claims that "all writing is atonement" (Ross, 2006), in her case, a reparative gesture of creativity born of angst and shame about earlier injurious behaviors and beliefs in relation to her mother. Jamaica Kincaid, in her memoir, *My Brother*, offers an account of the death of her brother, Devon, of AIDS on Kincaid's home island of Antigua, to which she returns for his pending death.

Of writing, she says, "When I was young, younger than I am now, I started to write about my own life and I came to see that this act saved my life. When I heard about my brother's illness and his dying, I knew, instinctively, that to understand it, or to make an attempt at understanding his dying, and not to die with him, I would write about it" (Kincaid, 1997, p. 196).

Kincaid goes on to explain that, when she was a teenager, her mother had burned her books "because I had neglected my brother when he was two years old and instead read a book" (p. 197). Implying a melancholic reparation, she speculates that "it would not be so strange if I spent the rest of my life trying to bring those books back to life by writing them again and again until they were perfect, unscathed by fire of any kind" (p. 198). Kincaid's memoir, while a self-professed effort to restore her own reading self, is also a reparative gesture to a brother whom she viewed with deep ambivalence, whose loss she mourns, and who, unlike her, did not have the possibilities created through migration, exile, and upward mobility in America, a difference that exacts on Kincaid a toll of shame.

There is a particularity to the relationship of writing and reparation within exile and migrant literature, perhaps because migration is, albeit not solely, such a profound (and ambivalent) experience of loss and displacement. The mythic link of exile and shame, guilt and punishment, borne of sin, resonates from the biblical Diaspora. Exile is a consequence of disobedience, a failure to adhere to the legislated prohibitions of a culture, a meaning Gatari Spivak claims has been largely erased (Spivak, 2000, p. 344). For queers, these historical meanings resonate deeply with an often self-imposed exile from a regressive home place—but home, nonetheless—that cannot offer a comfortable zone for emerging queer identities and, because queer identities go against established norms, that often can promote shame and self-loathing at the expense of healthy emergence.

O'Toole represents his own ambivalent structure of feeling often associated with leaving, questioning while forging a relationship between loss and writing: "Oh—I was eager to go. Yet, ironically, there was a far greater pleasure in coming back and being reminded of my first aches, as strong as an adolescent's sexual urges, to leave. Perhaps my destiny in leaving was to build a house of regret for doing so, which I would later dismantle, down to the last rusty nail. And then I would sit down and write about it" (p. 70).

This reparative gesture—writing to assuage regret—appears to be heightened in queer accounts of migration, such as O'Toole's. O'Toole's memoir, which captures the crisis of deaths in North America's gay communities, and coinciding, as it does, with massive moratorium-related outward migration and the creation of "ghost ports" (Walden, 2001), cuts an emotionally compelling line to connect the trauma of multilayered losses related to security, belonging, and multiple meanings of home—*here* and *there*, to echo Huston (2002). Calling AIDS "the most nightmarish of all the endless night we endure," O'Toole says, "It is the very end of home" (p. 111). Bringing together these losses—through migration, difference, trauma—O'Toole effects a narrative closure that demonstrates the processes of mourning to which Klein refers while also demonstrating that crisis and loss can be the impetus to recreate "villages" (whether the gay community or outport Newfoundland) in ways more compelling and vital than their previous versions.

As might befit a pilgrim's memoir, O'Toole's story ends at the Camino, *Camino de Santiago*,[12] a place that, through its ocean, people, cultural practices, and climate, shares an affinity with Newfoundland and Labrador. Here, like many pilgrims before him, O'Toole has an extraordinary spiritual experience through which, for a brief moment, he is united with his lost loved objects:

> Something took hold of me on that drive [the pilgrim's route]....I began to weep softly, and this soft weeping would not cease.... I began to think of all the pilgrims who had trod this road before me and I felt overwhelmed by their proximity to me....I was extremely close to something I had never neared during my lifetime.... All these people... souls...filled every available space of the landscape and beyond. I could *feel* them [those known, loved and lost, those unknown and also lost] all around me....They were all here for this time that I wept. They were not gone, they had returned for a brief while. (pp. 198–99)

In this moment, O'Toole experiences the profundity of love and forgiveness, the beginning of peace as "the end of expectation—the end of longing" (p. 199), and *mon joie*, the ultimate moment of acceptance, forgiveness and reclamation: "I'm gay, I'm a recovering alcoholic, and I'm from Newfoundland. I could not have *asked* for a more interesting life" (p. 199)—Klein's reinstatement of goodness through effective mourning.

At the Camino, this moment of peace and insight, avowal and reclamation, and of feeling a part of a larger whole, is O'Toole's reprieve from "the end result of writing [that] leaves the writer stranded, panicked, and alone" (p. 194). If, as Holly (1999) contends, "writing never heals" (p. 14), what writing can do is bear witness and bequeath. And it can urge the creation of new meanings, new sites of attachment through the acknowledgment and acceptance of loss. In this sense, then, writing is both reparative and transgressive, overcoming the limits of loss through squarely facing them so that they may reveal ways forward.

Finding Loss: The Irish Connection[13]

The first Irish in the New World were indentured servants in Newfoundland. For hundreds of years, Irish have immigrated to Newfoundland to fish. The Irish name for Newfoundland is *Talamh an ÉÉisc*, "the land of fish." Newfoundland is the only place outside of Ireland the name of which is uniquely represented in the Irish language.

The cultural affinity of Ireland and Newfoundland is reflected in many and diverse ways, an everyday reminder of how culture "generates" across time, space, and oceans. Cultural legacies, as O'Toole argues and demonstrates, can be affective as well as material. O'Toole's memoir, located as it is, in large part, in a remembered and idealized community inhabited mainly by Irish immigrants, mostly Roman Catholic, and many of whom fled The Irish Famine,[14] is a testimony to the "emotional history" (p. 67) that O'Toole argues is a large part of the Irish legacy in Newfoundland. O'Toole notes that "the influence of the Irish and their Roman Catholicism—the highly musical language, alcoholism, low self-esteem, fortitude, willfulness, natures that are a seductive blend of sweetness and dark despair—has been the strongest on the island" (p. 67). Much of what O'Toole describes as the character of his "endless night" as a Newfoundlander he traces to this emotional history and its continued resonance in individuals and the culture as a whole.

O'Toole's ancestry can be traced to the Irish immigrants who desperately fled the Famine, a fact that adds a weight of history to his oft-repeated metaphor of "hunger" and longing. "My mind always felt like an empty stomach" (p. 72), O'Toole says of his longing to know, to experience other places and people: "Without the hunger, I would have gone nowhere. My mistake was thinking I had to abandon my heritage to do it" (p. 73). On another occasion, he comments, "Whatever is essentially worthwhile about me comes from Renews, from my first need to escape it and from my desire to return to it. It cast my dreams for me; its deprivations made me hunger" (p. 42). Those deprivations—or the hunger to which they gave rise—are what O'Toole misses about Heart's Longing, and what feeds his most nostalgic moments. "I love a longing for it," he comments, and he also loves the hope longing engenders (p. 146). O'Toole acknowledges that longing can be a state of "limbo" unless one acts on what it is for which one longs. Otherwise, longing is melancholic petrification, a characteristic he also quickly attributes to both Newfoundland and Ireland: "The expectation of loss, this underlying dread, has somehow, over centuries in the Irish and their descendants in Newfoundland, been encoded through a process too mysterious, yet all too ineradicable, for words" (p. 59). O'Toole aligns loss, lack, and longing: "To not have is to look forward to having....Longing, for the Irish, for Newfoundlanders, can, like melancholy, be an end in itself" (p. 145). In the memoir, it is the movement through melancholy to mourning that lifts the limbo of longing and resists the trap of this "end in itself."

Much has been written about Ireland and the persistent metaphor of Ireland as (feminine) child. Richard Haslam (1999) provides an overview of the literature that represents Ireland as a "damaged child." Calling the metaphor "a manifestation of the pathos" of melancholy, he argues for new metaphors to capture the diversity and complexity of Ireland and "the Irish." Likewise, O'Toole employs the same metaphor for Newfoundland when he describes its history and its current place as a province within the Canadian nation. This image of the abject, forlorn, and desperate child is reiterative, found in historical and contemporary representations of Newfoundland and Labrador, of which O'Toole's is but one. "These people [of Newfoundland and Labrador], the majority of them of Irish-Catholic descent who fled the great potato famine on the Auld Sod, have pretty much always been poor and, until quite recently, uneducated in the sense of readin', 'ritin'n' 'rithmetic.' A British colony until 1949 and sucking the British Empire's hind teat until that time, Newfoundland became a Canadian province and then moved to Canada's hind teat. It's the place nobody wants, an unwanted child" (p. 2).

O'Toole's own initial description of the place is saturated with abject imagery—the often (re)cited views of the colonizer enshrined grotesquely on the lips of the colonized through hyperbolic and negative comments that range from the quality of cultural character to degree of progress in modernity.

Conclusion

Such is the nature of being alive: the passage of things, their eventual loss, and how you must accept them if you are to go on with your life.

Lawrence O'Toole, *Heart's Longing*

The tone and sentiment of O'Toole's memoir, in the end, contrast sharply with its beginning, capturing the emotive shifts of the psychic process of mourning. Through the reparative gesture of writing, O'Toole transforms his relationship to place, identity, and culture in ways that enable more than inhibit. Reconciliation and the beginnings of inner peace now lessen the shadows cast by the turmoil of "endless nights." Wisdom and tenderness replace self-loathing and arrogance, the latter what O'Toole had called "the most secure form of self-defense" and "a part of the low self-esteem I carried around as a Newfoundlander" (p. 105). Arrogance is an attribute commonly found in the migrant expatriate, for whom it offers a pretense of superiority against a simmering, unresolved loss. Echoing Alice Miller's assertion of "a new empathy with [one's] own fate, born out of mourning" (Miller, 1997, p. 15), O'Toole counters arrogance with a growing humility, sincerity, and love, thus finding home where it can only ever be—within oneself.

As part of my wider discussion of the questions that informed this paper, I argue that cultural narratives such as O'Toole's memoir are needed in these times of cultural crisis, change, and loss, when what is untenable and unmourned must be appropriately let go and in its place forged new attachments and new belongings—to people, places, objects, and ideals. Spivak (2000, p. 353) contends, "We who are from subordinated cultures must accept the new as agents as we mourn the past with appropriate rites as subjects." O'Toole's memoir offers an opportunity to consider, in cultural terms and in multiple ways, a longing to be "a subject of loving rather than, at best, an object of benevolence" (Spivak, 2000, p. 352). It is also a textual site on which to reconsider the viability of communities faced with great devastation and loss and the conditions under which they might be reconstituted. O'Toole shows how crisis wreaked havoc on both his beloved villages—in New York and Newfoundland. The epidemic of HIV and AIDS ushered loss through gay communities worldwide. The fire that spread its hot breath of loss through Heart's Longing portents, perhaps, a later, larger loss as the cod fishery ends, decimating hundreds of communities. But, O'Toole also shows how each of his beloved villages survived by facing its losses, building on its collective strengths, and, through caring, collective struggle, by being a village, in the best sense of the word. In this way, O'Toole's memoir may be a parable for our times, one that attests to the life-giving properties of loss confronted and that presents renewed joy and love (re)borne of a hard settling of our scores with sorrow and loss.

More generally, more such cultural representations are needed that capture the complexity of vectors of identity as diverse, fluid, and contradictory positionings made and remade through an ongoing act of "becoming," not "being," and that address the cultural problems and possibilities of melancholic disavowals and positions. Such narratives could choreograph and stage an improvisational, forward-looking identity dance, a lithe stepping, critically attuned with the past, the present, and the future—a politically progressive identity reel. The challenges inherent in the changes afoot culturally, in local and global contexts, require a collective and educational project of cultural renewal and change sensitive to both the legacies of struggle and difference that shape a place and to the processes by which their melancholic shackles may be understood and, also, loosened. A commitment is needed both to write and to read such narratives as part of a new and emerging cultural project and as part of a body of complex, honest, and forwarding looking narratives. A commitment is also needed to educate through and with such narratives as a way to address what is before us, culturally, and to invite a dialogue about loss and hurt and their ever-presence in our educational spaces. Such dialogical spaces help us, culturally, to find our grief and to mourn, heal, grow, and change as part of "learning to love again" (Brown, 2006)—differently.

POSTSCRIPTUM: PASSING

The bus charges through the countryside of Northern Ireland, along narrow roads randomly marked by small stone houses dotting rolling green fields, through County Tyrone, one of the counties that, in the Republic, are referred to as "the lost 6." I try to embrace what it is I am seeing and feeling as I anticipate my arrival in Derry. When I was a child, Bernadette Devlin had cast Derry forever in my imagination, her forceful, impassioned addresses regarding "the troubles" an early and rare televised image of a strong, gutsy, defiant, yet somehow also vulnerable woman determined to fight for what she believed. Now, forty years later, it is the eve of an historic agreement to legislate an end to these historic troubles. There are no longer any British Army checkpoints in the North, but the persistence of graffiti reminds of how fiercely and deadly the lines were drawn. My thoughts are interrupted as, passing through the village of Clady, I notice small cars forming long lines on both sides of the road, their epicenter a small house set back only slightly from the road. A blur of people attired exclusively in black and white provide enough detail. It was a wake, "old country" style—still held at home, where life and death embrace most intimately.

I consider this happenstance resonance of my purpose here in Ireland, a part of my own search to understand so as to think and to write meaningfully of my own place, changing under the weight of different sorts of troubles, and a place so deeply linked, historically, to this one in which I now travel. I think of those lost through the Famine and the troubles, who have been uncannily present since I arrived in Ireland, as well as those lost to the ravages of disease and war and indifference seemingly everywhere. By now well past "the funeral home," I am drawn back to another continent, another story, another home in another village…and two sisters, lost in their grief, each thinking how queer it is they cannot find it.

CHAPTER 3

"THE WORD, FOR LOSS"
LITERACY, LONGING, AND BELONGING

INTRODUCTION

Against a backdrop of public discourse that reduces literacy to abstracted skills and abilities, largely outside—or only with glib consideration—of social relations, and in which international literacy test scores rank nations and peoples on a global scale of progress in (post)modernity, too little attention continues to be given the sociocultural nature of literacy. New Literacy Studies[1] scholars have insisted, for decades now, on the situated nature of literacy (Gee, 2001; Street, 2001) through their familiar refrain of "literacy as social practice," but teachers continue to be guided by testing schedules and demands to enhance test scores, sometimes under threat of job security. Under such prevailing conditions, and despite a proliferation of research that directs us otherwise, the insights of New Literacy Studies remain marginalized within both public discourse and public school practices. In Canada, the province of Newfoundland and Labrador regularly scores lower than many other provinces on nationwide standardized tests. While New Literacy scholars would see their problematic character, such scores, usually presented in decontextualized ways to feed a sense of a "literacy crisis," hold great public sway, both grabbing and reinscribing the cultural imagination in particular and often deeply conservative ways.

The political dimensions of these conservative statistical dramas may be obvious to New Literacy Studies scholars, but less obvious are the psychic dimensions they can also register. There exists a dearth of research that might provide more compelling insight into the psychic complexities of literacy learning. If, as New Literacy Studies scholars argue, literacy is situated, social and political, then what emotional dynamics might operate within the deep particularity of specific social contexts to mediate relations to literacy technologies? This chapter is a partial exploration of the psychosocial aspects of literacy. Focusing on the *structures of feeling* (Williams, 1973) that might accompany some prevailing and persistent beliefs about literacy, the paper highlights the complex intersections of history, memory, culture, and identity, and how their relational character may come to bear on longings for, beliefs about, and cultural practices of literacy. Framed around an analysis of *Ann and Seamus*, written by Kevin Major and illustrated by David Blackwood, this chapter also offers something by way of what psychoanalytic inquiry might bring to literacy research and, in particular, to New Literacy Studies. Specifically, I use this discussion to inquire into the nature of certain attachments and their intersection with projects of literacy as they relate to social and cultural transformation, with particular reference to contemporary Newfoundland and Labrador.

History's Contemporary Resonance: *Ann and Seamus*

Ann and Seamus, a verse novel,[2] is a poetic rendition of the story of Ann Harvey, a largely unsung hero(ine) of early nineteenth century Newfoundland. Ann and her family, at that time in history the sole inhabitants of Isle aux Morts, a tiny community on the southwest coast, twice saved large numbers of seafarers shipwrecked on the reefs and sunkers that form the coastline of Newfoundland and Labrador. The rescue described in *Ann and Seamus* of 163 Irish emigrants aboard the British brig *Despatch* out of Londonderry, Ireland, en route to Quebec City in 1828, earned the family a gold medal from the Royal Humane Society. By all accounts, it was seventeen-year-old Ann

Harvey's bravery that was central to the rescue effort, and, in recognition of her efforts, George Harvey, her father, gave her the medal upon its presentation to him. Published in 2003, the verse novel is a stirring fictionalized account of this little-known event and the figures central to it.

Ostensibly a story of heroism and fleeting romance, *Ann and Seamus* is also a provocative study of the vicissitudes of place, longing, and belonging—and the place of literacy in the emotional geography that underscores the narrative. In his narrative of Ann Harvey, Kevin Major chooses to layer the story of heroism with a fictional story of romance between Ann and one of the shipwrecked Irish emigrants, Seamus Ryan. In so doing, and by giving this romantic plot the central status through the choice of title, Major attempts to appeal to a widespread audience, appearing to offer something to both young male and female readers. The narrative is comprised of three separate sections, each named for the character whose point of view it represents. Ann's voice opens and closes the narrative, with the voice of Seamus wedged between. Despite the suggestion of the title, romantic longing is surpassed by another, more profound one—a longing to be free. Most of all, Ann longs to read and to write, which she equates with freedom from Isle aux Morts. Seamus longs to be free of English tyranny, to live a self-determined life far from servitude, which means, in his case, far from Ireland, as well. These tensions—of belonging and freedom, of love and loyalty, and of desire, dreams, and self efficacy—form the rich texture of this short narrative.

Ann's longing to learn to read and to write is identified early in the story and remains a central feature of the narrative structure until its very end. As a narrative focus, however, it has been ignored by reviewers and readers. This oversight is in keeping with the unquestioned status of literacy: just as literacy is a naturalized *demand* of contemporary society, it is also a *naturalized* element of the story, positioned as a taken-for-granted feature rather than a provocative and *motivated* authorial choice. I want to highlight the importance of locating authorial choices within a broader cultural context and consciousness, troubling these choices, and assessing their sociocultural implications. Examining the ideological fabric of literacy (Street, 2001) through such textual representations as that of *Ann and Seamus* also reveals some of the discourses that circulate and position us, as cultural subjects, in complex, often confining ways. While *Ann and Seamus* fictionally reenacts a two-hundred-year-old story, it does so through discursive frames that are very contemporary and, as such, potentially revealing. Since its publication, the story has gained much prominence, as may be evidenced not just in book sales and translations, but in adaptations, too. Of particular note is the operatic adaptation by the award-winning composer Stephen Hatfield for *Shallaway*, a St. John's based and internationally acclaimed choir directed by Susan Dyer Knight.[3] Entitled *Ann and Seamus: A Chamber Opera*, it premiered, prior to a world tour, in July of 2006 in St. John's, the provincial capital, with several descendants of Ann Harvey in attendance. The opera, like the book, has traveled extensively, gaining cultural momentum and recognition as it goes. Cultural texts intersect in complex ways with the contexts in which they are taken up and "read." Such textual prominence, then, begs the question of why a story of a brave young woman living with her family nearly two hundred years ago on a desolate coast of Newfoundland might resonate in contemporary global culture and, in particular, in contemporary Newfoundland and Labrador culture. An examination of Ann's longings, as depicted in *Ann and Seamus*, may reveal something of the nature of this contemporary resonance, the structures of feeling that can surround longing and attachment, and their force and influence in the broader repertoire of social relations. Such examination may also provide insights into the lives of many contemporary Newfoundland and Labrador youth—and, perhaps, many others in similar contexts worldwide—whose dilemmas bear some resemblance to those of Ann, despite the nearly two hundred years that separate them.

In 1828, the time of the story told in *Ann and Seamus*, Ann Harvey is seventeen years old, the very age at which most contemporary youth of Newfoundland and Labrador—and elsewhere—are poised to move forward from public school into young adult life. Unlike these contemporary youth, Ann lives in an early nineteenth-century world where alphabetic literacy is a personal dream, not a cultural demand. Yet, perhaps very much like them, Ann's mind is full of questions that reveal the nature of her dreams, her longings, and, as well, her anxieties. But, through her questions, her critical acumen—the already embodied pose that is the desired object of critical literacy—is also seen. In the novel, Ann poses a dozen provocative questions. These questions do more than structure and move the narrative along; they more importantly suggest the desires out of which the questions are formed and the psychic struggles that infuse Ann's contingent responses. These questions, many of which are part of inner monologues, reveal much about this fictionalized Ann Harvey—her astuteness, her emotional attentiveness, her worldliness—and her rich inner life despite the minimalist culture in which it develops.

In the first section of the novel, Ann's questions come out of the social organization of her life on Isle Aux Morts. Her inquiry touches many aspects: of her home place—"Isle Aux Morts. / Who would ever want it said / they dwell in a place with such a name?" (p. 11); of her mother's burdened life of child bearing and raising, house keeping and cleaning, on top of her role in the family fishery—"Is this the world my future holds for me?" (p. 17); of the fishery—"This, the reason we are on this earth—/ to turn cod into dried salt cod / for the tables of the world?" (p. 18); on the injustices of educational access—"But the fish merchant's sons have their learning / or how would they grow / to be merchants like their father? / Their daughters the same. / They have their books." (p. 22); on her greatest wish—"Mother, one day shall I read books?" (p. 24); on dreaming of a life beyond her present confines—"Tom, do you ever think of what's beyond the sea? / ... / Do you know that some your age have never seen / the sea?" (p. 26); and on her destiny—"Will I forever dance in the misfortunes of others?" (p. 31). In the final section of the verse novel, when, as a result of her romance with Seamus, Ann is faced with the real prospect of change and departure, the nature of her questions change and they become less wistful and more anxious, indicative of the turmoil she feels: of romantic love, she asks, "What stirs me so / what turns my lips / to return his smile / what beckons me away / to where no one / will know" (p. 97); on a future with Seamus—"But Seamus, what is the life we'll have?"

(p. 96); and of the decision she must make—"My head turns in circles / and my heart / does not rest. / Will it ever?" (p. 99).

A heartbreaking exchange of questions between Ann and Seamus marks the climax of the novel. Seamus pleads, "Ann, come away." Ann responds, "Seamus, what am I to do? / Leave for a far-off world perhaps worse than my own?" Seamus wonders, "But Ann, your dreams—/ do you not gaze out to sea / and yearn for what is beyond?" Ann counters, "Seamus, you have nothing, and I nothing but the solid rock of home." Desperate, Seamus pleads, "Ann, do you not love me?" (p. 105). Perhaps more pragmatist than dreamer, after all, Ann's response to Seamus is her final question: "Will it fill a lifetime?" (p. 106). If it can be argued that life is about questions of attachment and separation—what to love and what to leave, what to hold on to, to keep, and what to resist, to let go—then *Ann and Seamus* offers an important opportunity to consider these issues within a specific cultural context and at a time when, within this context, such questions—of what can, does, could, or should fill a lifetime, and the capacity and freedom to choose—are ever present.

Longing to Learn; Learning to Be Free

The longing for freedom that Ann and Seamus share was nurtured in the particular confinements of soils thousands of sea miles apart. In Ann's family, her mother and she bear the burden of such unrequited longing, dreaming, and questioning. Ann's mother, Jane, has learned to silence her longing (a lesson in which Ann is actively schooled throughout the story). Ann says: "My mother Jane has too much / filling her days / to ever think / of singing" (p. 24). Still, Ann reads her mother's muted expressions of hope: "My mother looks at me / and knows what questions dwell within my head / though she hardly has the time / to be giving me answers. / In her anxious eyes I see / someone who wishes more for her daughter / than she herself will ever know" (p. 24). Notably, even if unintentionally, Jane is also the only character in the narrative unrepresented in Blackwood's etchings. Invisible and on the margins of the narrative, her withdrawal and resignation are in marked contrast to the contentment suggested in the animated demeanors of George, Ann's father, and Tom, her brother.

The men of the Harvey family refuse to question their lot. As they both see it, dreaming is for idlers. Ann differs. Early in the story, when Tom spits an accusation of "idler" at Ann, she counters insistently, "dreamer" (p. 21), as if to suggest that there is nothing idle or nonagenic about dreams. Her father chides her: "No time for discontent, Ann, me young maid. / There be lots worse off, me love, / lots worse off than we" (p. 20). Still insistent, Ann replies, "Lots better off, too, / if it be told. / God in heaven has his ways, I grant" (p. 30). Ann grows silent at her mother's reprimand—"Don't be questioning the Lord"—and her father's reminder—"Thank the good Lord for the sense / to know what ye're about" (p. 30). This notion of (good) sense would be summoned later, by Ann's mother, in the same regulatory manner of suppressing questions and diffusing resistance, when she attempts to persuade Ann to refuse the lure of romantic love and to stay in Isle aux Morts. The unquestioning stance of the Harvey men is part of the problematic and deeply gendered mythology of life on the sea to which so many are so strongly attached, despite its hardships and, historically, its meager financial rewards. In his memoir, *Heart's Longing*, discussed in the previous chapter, Lawrence O'Toole (1994) explains the allure of this mythology:

> Out on the ocean [the fisherman] is untouched by the common cares of the land, cradled instead by the deep, caressing roll of the waves. The sea rolls and dips and carries the small craft up and down liquid hills. A dream with the eyes wide open. And the man can become a child again....As the fishermen bait hooks or haul trawls...a look of utter contentment of which they are not aware...passes over their sea-braced faces. They cannot and do not ask for more, having found their place in the world, as difficult, bone-tiring and devoid of tangible reward as it manifestly is. This is essentially where these men want to be, and it is where they are. (pp. 16–17)

With the introduction of Seamus Ryan, whose passion for freedom is fueled doubly by an anger at the British imperialists and a loyalty to his extended Irish family, Major introduces us to a strong image of male longing. Seamus is unfazed by uncertainty, driven as he is by a "thirst for a better life" (p. 49). Seamus is recognized as one who can "make something of himself" and "make way for the others" to follow (p. 47). As an Irish rebel, Seamus is "stoked in the fires of freedom" and, yet, as he asserts, his desire is to be "free of the feuding" (p. 46). The same British imperialists Seamus abhors are mirrored in the faces of the fishing merchants who exploit Newfoundland fisher families such as the Harveys. Both Ann and Seamus are bound by these attachments forged, albeit differently, through family, religion, and nation. Each dreams and each sees a place for literacy in these dreams. Like Ann, Seamus believes that literacy is intricately tied to freedom and to outward and upward mobility. Seamus has seized literacy, stolen it from under the noses of the British imperialists at one of the hedge schools. "I am the eldest son sent ahead / because I am the strongest / and at the hedge school / we built behind their English backs / the priest taught me to read. / I will make way for the others" (p. 47). To secure another of his longings, "[to] find meself a girl at that" (p. 49), Seamus promises to bestow his bounty of literacy onto Ann and he uses it as a lure for their future together: "We'd make a new world... / I will teach you the ways of books, he says, / to recite the wisdom of the world, / to become a wonder to behold / in dresses more dazzling than the sun" (p. 92).

Ann is thus perched on the cusp of learning and leaving. Sensing her reticence, Seamus asks, "Ann, do you not love me?" (p. 105). Her doubtful question, "Will it fill a lifetime?" (p. 106), is, in the end, also her answer. In the end, Ann sides with doubt and decides to stay and, in so doing, she breaks rank with her dream of literacy founded on leaving. The unbearable heartache she would cause her family, the anxiety of leaving the solid rock of her Newfoundland home, and the risks of the unknown form a nexus of refusal. The *untested* promise of her dreams—and of literacy itself—fails to withstand the (albeit incomplete) fulfillment of home. Bound by loyalty to family and place echoed through conservative refrains of religion and nation, the fictionalized Ann succumbs to their *demands*. For Ann, deciding to stay is deciding to not learn to read or to write. The real Ann Harvey would die in 1868, decades before the small Church of England school was built on Isle aux Morts at the end of the century, and longer still before compulsory schooling was introduced to Newfoundland.

Why would someone who so desperately wanted to learn to read and write turn their back on its opportunity? Why would a young woman who so fearlessly faced the raging seas off Isle aux Morts *fear* leaving? What happens to Ann in the spaces between her dreams of learning to read and write and her decision not to accompany Seamus, to whom she is attracted and through whom it appears she might access this dream? As one reviewer notes, "because the real-life Ann stayed in Newfoundland and married a local man, Major must attempt to explain her choice not to join the dashing Seamus. He has already established his heroine's character as stouthearted and yearning for adventure, so readers may question her decision to remain in her monotonous life" (Ortega, 2005, p. 168). The gap is closed on such hermeneutic dissonance, but of interest here is the manner in which closure is effected and the ideological considerations that arise from it.

Under closer scrutiny, Seamus would seem an unlikely literacy *sponsor*[4] (Brandt, 2001, p. 557), and, while the inadequacy or total absence of viable sponsors bears heavily on the realization of Ann's dreams, it does so in ways not typically found in the literature associated with literacy sponsorship. Ann's emerging relationship with literacy demonstrates the deep-seated structures of feeling that are part of what it means to navigate the difficult emotional terrain of learning. What Ann lacks is a certain kind of sponsor, one who can *bear* her longings and the possibility of her leaving that they bring. Ann is able to dream of being literate, but she cannot imagine herself a literate person. Ann has no supportive networks that might make her dreams bearable or fully imaginable—as a part of her identity kit (Gee, 2001, pp. 526–27)—ones to help her hold and manage the anxieties that are borne of the impending loss posed by realized dreams. Ordered to suppress her questions, told her dreams are idle and lacking sense, and encouraged to allow fear to guide her and the status quo to comfort her, she recedes from the direction of her longings. Ann is restricted to the insights and alliances formed in her primary discourse—those that discourage change, recoil from difference, and demand an acceptance, not a questioning, of one's place. For Ann, her primary, fundamentally conservative and limiting discourse provides a glimpse of a dream beyond the shores but no means to understand the (im)possibility of traversing the distance.

Can Ann herself bear the separation that leaving (and learning) would entail? While it is easy to imagine the displacement that Ann might feel upon leaving her close-knit family, it is important to see this separation she fears as more layered and complex. Early in the story, Ann muses, "I swirl free in my dreams / a head full of notions / of other places / and other times / swirling now / whirling in the wind / and where will I come to rest / the question drifting about my waking hours / lurking through my nights" (p. 32). Here, her freedom is modified by an anxiety about where her dreams might take her, an anxiety that continues to haunt Ann as she ruminates on the dilemma her relationship with Seamus poses.

While learning to read and write may offer new possibilities, as Gail Boldt points out, "these changes often simultaneously involve the loss of an easier communion with the people and ideas to which we once turned for love and reassurance" (2006, p. 286). The compounded effect of the losses posed by leaving and learning may make either or both unappealing. By choosing not to leave, Ann resolves anxieties around each of these possibilities. Seamus,[5] the supplanter, fails to displace the loyal Ann. Consider the structure of emotion that informs Ann's question as she considers leaving with Seamus: "The song I hear / is filled with doubt, / with an aching, heavy-hearted fear. / What of myself in days to come / without a talent to call my own? / What of myself without the sea, / unable to read or write of home?" (p. 100). Note that Major writes here, "*Of* home," not *to* home. Without a place in history, without her beloved place a place in written history, without the means to read and to write *of* and about what she loves—to find solace in cultural objects of literacy—her future is unfathomable, her memories seemingly inaccessible. Of such anxious awarenesses, Gail Boldt writes, "Because reading is a powerful discursive presence in local and national politics and group identity, students may in fact be demonstrating more wisdom than we have been willing to acknowledge in their suspicions about the unproblematic efficacy of reading in their lives. I am arguing, quite simply, that we respect the possibility that our students know something about the dangers of how learning to read may negatively interfere with their lives" (p. 301).

The Graceful Dream of Literacy

Ann,[6] whose name denotes graciousness, dreams of a literacy that reflects Sylvia Scribner's (1986) metaphor of literacy as a state-of-grace, one in which literacy is aligned with values such as social mobility, civility, intelligence, worldliness, respect, and an alliance with elite culture. Despite her astute observations on the injustices of the fishery and her own lack of access to learning—"But the fish merchant's sons have their learning / or how would they grow / to be merchants like their father? / Their daughters the same. / They have their books." (p. 22)—Ann cannot recognize the political parameters of literacy's history. Ann does not recognize the ideology of the colonizer that her dream of literacy as state-of-grace embodies. Nor can she recognize the manner in which the object of her desires—literacy—constructs her as abject, as "other," as outside the parameters of its privilege. As Casmier-Paz (2000) writes, "Literacy discourse functions primarily to locate underprivilege, and assert the value of written language…A "literacy crisis" is possible because literacy discourse does not recognize that ontological issues related to writing and economic / social uplift do not inhere in the relationship of subjects to written language. Again, the crisis is not one of literacy but of *legibility and legitimization*, when literacy discourse narrows the range of signifying practices that are deemed valuable. Our understanding of literacy is the product of a history which has reduced human consciousness to that which is produced by and for written language" (p. 8).

The verse novel *Ann and Seamus* draws on and mobilizes historic binaries to sediment this power of literacy in ways that continue to mark our current debates and discussions about literacy. The narrative is structured on many explicit "binaries" that are familiar to the cultural context of Newfoundland and

Labrador both historically and today: of nation—the English and the Irish; of religion—the Protestants and the Catholics; and, of social class—the merchants and the fishers. However, my focus here is on those binaries more implicit to the narrative and that (continue to) locate Ann—and her contemporary compatriots—as poor, isolated but content, as lacking agency, and as resisting modernity. These implicit narrative binaries include: literacy / oracy; writing / speech; consciousness / memory; civility / primitivism; history / mythology; privilege / marginalization; modernity / premodernity; empire / colony; intellect / emotion; abstractionism / pragmatism; progress / stagnation; autonomy / dependence. These binaries align to render comprehensible another story, one the traces of which continue to haunt efforts in Newfoundland and Labrador to move forward culturally.

Ann's family do not read or write alphabetic texts. Theirs is largely an oral culture, based in speech and utterance and remembered and retold in story and song. As Casmier-Paz (2000) reminds us, literacy discourses assert and reinscribe the value of *written language*, of writing and reading. Positioned outside such literacy discourses, Ann and her family are rendered "primitive," lacking the blessings and graces of alphabetic literacy. As primitives, they represent all that modernity rejects—what would appear to be the more likely source of any cultural "shame" in poor literacy scores in contemporary contexts. Within modernity, to lack literacy is unthinkable. The final line of *The ABC Song* is a reminder of *litteratus* (Latin, "acquainted with the letters of the alphabet") from which the word *literacy* is derived, and how it functions to separate, to create one as a thinking and thinkable subject: *Now I know my ABC's, tell me what you think of me*. Without alphabetic literacy, the subject is unthinkable, unthinking and unvalued—she cannot be thought of because she does not exist in history, in literacy discourse. At the end, as Ann and Seamus fade into each others horizons, it is Ann who fades into history, forgotten, unwritten, a partially enunciated figure within a cultural myth, without history—without literacy.

Yet, in *Ann and Seamus*, the cultural ledger sheet drawn up by an acritical acceptance of the value base inherent in such binaries is at least partially questioned and challenged as we watch Ann navigate some of the potential gains and losses offered through literacy. For example, counting the number of saved and lost from the *Despatch*, Ann says, "Despite what the merchant man would wish / we know the ways of numbers, / the price of their misfortune" (p. 88). That these binaries are partially deconstructed may do more to reinshrine their emotional structure—for their power has not been fully disclosed—than it does to dislodge the structures of cultural power to which they attest. There is, then, a real danger in such textual deployment of continued reification, a danger apparently realized through the particular and naturalized literacy discourse that frames the story.

The Word, For Loss

In "Mourning and Melancholia," Freud (1917/1989) distinguishes these two grievous responses to loss. He writes that mourning is "the reaction of the loss of a loved person, or to the loss of some abstraction which has taken the place of one, such as one's country, liberty, an ideal, and so on" (p. 586). The psychic process of mourning or "letting go" succeeds when energy is slowly but fully withdrawn from the lost object and the mourner is able to invest in new objects. In a state of melancholia, however, Freud argues that the mourner denies loss, and, instead, reinstates the lost object within the ego in order "to establish an identification of the ego with the abandoned object" (p. 588). Anne Anlin Cheng (2000) notes that "melancholia alludes not to loss per se but to the entangled relationship with loss." (p. 8). Grief is deepened in melancholia because loss is disavowed or is not fully acknowledged. The melancholic ego constituted through such a response to loss is, in a manner, a compilation or an incorporation of its losses, an identity contingent on its relationship to absents.

Melancholia, as a form of disavowal and a devotion to lost objects, inhibits the formation of new attachments, more diversified identities, and new futures. The melancholic lives in *denial of self* in exchange for refusing unbearable loss. Ambivalence, a love-hate relationship with what is lost, is, not surprisingly, a characteristic effect of melancholia because the melancholic both needs and resents—loves and hates—that which is its constitutive base. Cheng (2000) argues that it is ambivalence that is the radical basis of melancholia as a critical tool. We can see some of this effect in the melancholic impatriate—which Ann may emerge as—or expatriate—Seamus's emotional lot in life?—who struggles to separate from and to confront ambivalence toward a place, a place loved because it is home but hated because it cannot sustain the kind of home one desires or needs. While heightened ambivalence may enhance critique and lead to change, diminished ambivalence tends to stabilize and work to maintain a status quo.

Contemporary scholars (Butler, 1997; Cheng, 2000; Eng & Kazanjian, 2003) recoup not only the interplay between mourning and melancholia but, more importantly, their coexistence and simultaneity, in individuals and cultures. Such critics see in melancholia a complex and *necessary* process of negotiation that entails a fierce determination to refuse the loss of precious social identities and the devaluing that often is attached to them. In cultural terms melancholia can be understood as a symptom of ongoing assimilation and its necessary losses and, simultaneously, a refusal of these losses, features that mark the cultural contexts out of which both Ann and Seamus emerge in history.

Literacy scholars increasingly are drawn to the implications of such psychoanalytic insights for literacy and learning (Salvio, 1998a, 1998b; Boldt, 2006). Walter Ong (2001) reminds that words are dead things, "the reduction of dynamic sound to quiescent space, the separation of the word from the living present" (p. 22), what Davis, Sumara, and Luce-Kapler (2000) evoke in their phrase, "breathless language" (p. 219). Preoccupied with word objects, literacy is a series of relationships to a signifying system of dead signs, a relationship to loss that prompted Julia Kristeva to argue that language is our melancholy burden (1980, p. 109). Ong developed an elaborate argument for the most generalizable effect of alphabetic writing, the first technology of the word: *separation*, noting, "[Writing] divides and distances, and it divides and distances all sorts of things in all sorts of ways" (p. 24). Yet, like all separations, those created by alphabetic writing technology also create opportunities for new and more manageable forms of intimacy (p. 30). This promise of new opportunity is the promise of literacy.

Like a melancholic person, *Ann and Seamus* incorporates many specters of loss. Significantly, in what is a very short narrative, Major uses the word *word* close to twenty times. *Word*, a concept in an alphabetic metalanguage of learning to read, denotes a print object—the alphabetized script of a spoken utterance. In an oral culture, it is misplaced (Davis et al., 2000, p. 216), so, Major's incessant usage is noteworthy. As a melancholic ingestion of the disavowed, its repetition functions as a return, a reiteration that calls attention to unresolved issues of literacy—both in relation to the cusp on which it locates Ann in her longing to learn to read and write and, it may be argued, in the broader cultural context of which the story is a part. Does Ann, through her refusal of literacy, refuse one form of separation—and subsequent melancholy—only to be swallowed by another, the emotional consequence of her disavowed dreams?

The Isle aux Morts—Island of the Dead—of Ann's day provides not just a backdrop for a pending sea tragedy but, more importantly, a psychic mise-en-scène for the animation of these dynamics. In the opening lines of the story, Ann questions: "Who would ever want it said / they dwell in a place with such a name?" (p. 11). Blackwood's art is central to the evocation of this melancholic effect. Drawn predominantly in shades of grey and blue, the latter color one that is synonymous with melancholy, David Blackwood's art is most fitting to this story. Known for his haunting, dreamlike etchings of outport Newfoundland and ennobled, mythologized depictions of its people, cast in the colors and tones of tragedy, Blackwood's images underscore nostalgia and the pain of melancholy. Despite her very active and key role in the rescue that is the impetus for the story, his choice to depict Ann as the only figure in the rescue boat in a passive position is a puzzling one. It betrays both an exclusionary consciousness *and*, more compellingly, Ann's petrified melancholic agency.

The bleak power and charged beauty of this backdrop shore up the complexity of a narrative about "dreaming beyond the sea," both in Ann's time and now, in contemporary Newfoundland and Labrador in the wake of the closure of the provincial cod fishery in 2003 The cod has iconic status in *Ann and Seamus*, as it has had in the colonial and provincial history of the Province. "Cod / is all the reason in the world / to settle in this cove. / Cod / fills our boat, / thick and lusty fish / some days in swarms as dense as fog. / Cod to gut and split abroad / and wash and salt and spread outdoors upon the rocks / to stack and store and spread again" (p. 18).

Major establishes the significance of the cod through its reiteration and stanza position (three consecutive times the word opens a verse alone on a single line) and, through imagery and increased stanza density (of lines and words) captures its powerful and abundant presence. But, despite this significance and abundance, Ann questions its centrality in the broader context of life's purpose and meaning: "This, the reason we are on this earth—/ to turn cod into dried salt cod / for the tables of the world?" (p. 18). It is a debate that continues two centuries later with a ravaged cod stock now considered "endangered," one that in the last three decades alone was fished to only three percent of its once great abundance. By dedicating the story of *Ann and Seamus* to "the families who fished cod off the shores of Newfoundland and Labrador," Major enshrines the story of Ann Harvey's bravery as a monument to the colossal loss that marks the end of the cod fishery. The fishery, the fisher families, and the once-abundant cod at the center of this cultural practice are lost objects and, in a very real way, Major's text may be seen as a provocation to move beyond melancholia and to properly and to fully mourn the way of life that is gone. In this sense, the verse novel may be seen as an invitation to dream life differently, to lovingly leave behind the lost object—and all that it represented—and to urge a productive use of creative energies to envision a new future. In this sense, Major's own words may be read as a reparative gesture toward what is lost, both in terms of Ann's story and the way of life of which, for over five hundred years, it was a part.

But can the story that is *Ann and Seamus* bear the weight of this profound gesture? Does its conservative inclinations override its fledgling critique? Or does it falter in its own melancholy, its own unreconciled and ambivalent representation—forged in a contemporary period of massive outward migration that exists against a historical backdrop of steady migration—of a culture's difficult relationship to leaving? Ray Barglow (1999) notes, "Despondency and despair correlate with what goes on in the social world....Historical circumstances can shape our lives as powerfully as the material bodies through which we live....In subtle ways, and along pathways that we might not have suspected, sorrow may be conveyed from one generation to the next" (p. 2). Barglow's point prompts an attention to the fictional Ann's losses, her sorrows. What might have become of Ann's dreams? In what ways might she have mourned her own loss of hope, love, and change? Did her broken dreams teach her silence, resignation, and inner rage, which appears to be the case with her mother? Who might be the contemporary bearers of Ann's unresolved sorrows? And how might contemporary versions of these historic sorrows, if still left unresolved, continue generational sorrow?

In *As Near to Heaven by Sea*, his popular history of Newfoundland and Labrador, published in 2001, two years prior to the publication of *Ann and Seamus*, Kevin Major presents a brief account of Ann Harvey and the selfless acts of her family that secured her and their place in the history of this province. With unabashed editorializing, Major writes,

> Today, few of us know of Ann Harvey. Many more extol the bravery of Grace Darling, the lighthouse keeper's daughter who ten years later, in 1838, help rescue nine people from a shipwreck off the coast of England. Wordsworth and Swinburne penned words in her honour. Victorian English hailed her in countless books and magazines. When she died they buried her in an elaborate tomb, built a museum, imbedded her story in the minds of every one of their schoolchildren. Except for the community tribute in Isle aux Morts, our sole public acknowledgment of the bravery of Ann Harvey was to paint her name on the bow of a coast guard vessel. We choose to neglect our heroes then wonder aloud if as a people we have an inferiority complex. Our ancestors should be lauded for their stamina in enduring life on a bleak, unwelcoming seacoast. And part of their story is the willingness, time and time again, to risk their lives for strangers. Imagine what would have been made of Ann if her formidable deed had taken place off Cape Cod? (Major, 2001, p. 193)

In this criticism of the absence of historical tribute to Ann Harvey, Major unintentionally beckons to the relationship of literacy and remembrance, to practices of signification as an evocation of, as well as a monument to, loss (Kristeva, 1980). But he is also unwittingly disclosing the very politics of literacy that shrouded the fictional Ann. Words, scripting, and reading—the markers of alphabetic literacy—were not available to many to pen and to hail the accomplishments of Ann Harvey in the ways Major wishes—and her colonial masters did not value the place or its inhabitants sufficiently to support them, let alone to pen their accomplishments. The accusation of neglect he so quickly inflicts on the collective "we" is more complex than Major allows and sounds unsettlingly similar to many of the contemporary *shaming* discourses of (self) blame and "being our own worst enemies," accompanied by calls for bootstrapism, that populate contemporary talk. Further compounding this "blame," in *Ann and Seamus*, through her refusal of literacy, Major implicates Ann in her own erasure from history. Such narrative moves, like our contemporary acritical societal and educational *demands* for literacy, may be viewed—at times and in certain circumstances—as more shame inducing than learning producing. Drawing on the work of Silvan Tompkins, Gail Boldt (2006) explains the relationship of shame and learning:

> The child, then, faced with the demand to learn something, may comply out of the hope of gaining love or of forestalling the loss of love. The child's act in the name of love is rebuffed every time she perceives, whether correctly or not, that her efforts are not good enough. Not only that, but the idea or the activity itself becomes a painful reminder of the failure or rebuff, and, even in the absence of negative human responses, the book, the math problem or the worksheet first associated with the shame provokes shame anew. The person, idea or thing that has the power to cause us to feel such shame then becomes the object of our anger and even sometimes our hatred. These affects can cause the child to withdraw from the source of danger, from the parent, the teacher, peers, and the curriculum. (pp. 290–91)

If contemporary test scores evoke shame, how might such shaming contribute to refusals of the project of literacy, and of schooling, more generally? Furthermore, if literacy separates, what powerful cocktails of threatened separation and evoked shame are offered youth struggling with the vicissitudes of attachment to place? Shame, self-doubt, fear, melancholia—the shameful exploitations of the colonizer have historically become, through cruel psychic twists, a legacy of shame for the colonized. The difficult maneuvers of love—in patriotism, nationalism and familialism—show love displaced onto objects sometimes better let go. This "call to freedom" is not freedom from attachment. But part of what learning must include is the caveat that some attachments—some efforts to love, to secure love—can make us decidedly unfree. Such caveats must guide us through our efforts at learning, at seeking connection, at forging new attachments.

Conclusion: Refusing Reading and Holding Fast[7]

In the end, in true melancholic fashion, Ann cannot bear the possibility of separation—and homelessness—that Seamus offers. Her mother's midwife, her father's fishermaid, and her brother's playmate—she is buried by attachments, from the chorus of which can be heard barely a whisper that urges her to dream, to be free. These stifling attachments coalesce in an idealized, romanticized, mythologized attachment to place, "the rock of Newfoundland" (p. 107), "the solid rock of home" (p. 105). Considered emotional pregivens, such attachments are negotiated as part of a field of action rendered commonsensical and contingent on disavowals, and fiercely resistant to questioning and change. In contemporary Newfoundland and Labrador, such melancholic urges continue to resonate, a cultural habit of affect, discourse, and representation.

The poet and writer Danielle Devereaux notes that, growing up in St. John's in the 1990s, following the 1992 cod moratorium, she was regularly reminded at home and at school that she was living in a culture that was dying and that required remembering and "saving" (Devereaux, 2006). It is important to ask of the nature of attachment formed in such an ethos. A proliferation of discourses of death and dying—as metaphors of meaning—continue to inhabit daily life in Newfoundland and Labrador and are distributed through media, education, and family, and are propped by yearly statistics of unrelenting outward migration.[8] These metaphors speak to and are part of the vicissitudes of mourning, markers of a culture grappling with colossal loss. Yet, these metaphors must be questioned, too, for the manner in which they enable or disable healing and promote or inhibit meaningful change. Many discourses can nurture a structure of feeling that is politically stifling and can be manifested in expressions both of unreflective migration and passionate impatriation. Each of these manifestations can represent a refusal of the difficulties and complexities of the context out of which they are borne—one in which there is both a cultural imperative to connect (or to remain connected), to *hold fast*, set against an economic imperative *to* separate, to leave, *to unmoor*.

Young people in Newfoundland and Labrador embody the strains of these demands of social identity. Unfettered patriotism is expressed in a determination by some to stay in a place at all costs, an obvious disavowal of self. Others exhibit the disavowing self-loathing that can characterize melancholy and talk of leaving, at first opportunity, a place believed to offer nothing, at worse, and to inhibit dreams, at best. But, there are others whose intuitions and actions show a more strategic resistance, who construct a direct relationship between educational attainment and leaving, in part what Michael Corbett (2007) captures in his phrase, "learning to leave," and decide neither education nor leaving is for them. For these young people, the pursuit of education is the willful mobilization of an inevitable departure from a place they are unsure they wish to leave. The refusal of education—and its central project of literacy—then becomes the refusal of a demand for change, with a return of loss and, simultaneously, a securing of home. Taking their cue from those who leave and tell of upward mobility but outsider status, they navigate the predicaments of economy and ideology in ways that are disturbing and revealing of the dangers of an unquestioned "holding fast."

In a similar, albeit fictional, irony, the absence of literacy that formed the impetus of Ann's dream to leave also makes it impossible for her to go. Just as migration presents loss of established belongings alongside the promise of new belongings, as Ong (2001) and others suggest, literacy—signification, more generally—allows us to access the creative force of loss, to inscribe, to analyze, to dream. But there is nothing free-flowing about such access or creativity. Bound by the political but imbued with the affective, literacy must be understood as both culturally specific and irreducibly personal, and lodged as deeply within the psychic as the social—the key tenets of any effective literacy project for these times, in this place. How these strains are represented within cultural narratives such as *Ann and Seamus* may suggest much about how we construct our pasts through present orientations that may or may not allow for better dreams and a more hopeful future of new belongings in many places, many homes. At a time of such real and pending social and cultural change, the need for new narratives of longing and belonging may well be what is in order. New versions of old narratives have their place in the transferences that are an inevitable feature of the fears and struggles of growth and change. But it is challenges to these established narratives that will unmoor us. It is new and disrupting narratives that may set us free.

Postscriptum: On Going

She would not have it that I stay. One does not succumb to fear, she would say. Neither does one resist it, resistance being yet another way that fear controls. Fear, she said, is to be embraced, held close in a loving gaze so to reveal its insights, bearing onto you an opportunity to grow. All her emotional thrust toward her children could be summoned in one word, *go*; all of mine, her daughter, in another word, *no*. Some nights she would come into my room and lay with me, reading aloud the old proverb on my wall poster at the end of the bed: *If you love something, set it free. If it comes back it is yours. If it does not, it never was.* We would laugh and talk.

I once asked, "So, love can't live with fear?"

"Not well," she had quickly replied. "But it would be a slow death."

On another night, I asked, "Why would someone come back who has been set free?"

"To appreciate that freedom can be anywhere, if you face your fears. Fear is your real prison," she had replied.

I persisted, "But what about those who do not come back?"

"Freedom can take a long time," she said. "Sometimes, more than a lifetime offers. But you have to try."

CHAPTER 4

SEPARATION, (RE)CONNECTION, AND A TRANSFORMATIVE EDUCATION OF PLACE[1]

I am not nostalgic. Belonging does not interest me. I had once thought it did. Until I examined the underpinnings.

Dionne Brand, *A Map to the Door of No Return*

In the Summer and Fall of 2006, my partner and I traveled to numerous small communities on the island of Newfoundland and the coastal portion of mainland Labrador. I had spent twenty years away from my home province, so part of my purpose in this travel was to revisit places in order to reflect on and renew a connection and, as part of a research project, to listen attentively and to bear witness to over two decades of change: To see a lay of the land, if you will, less through the nostalgia-leaning attachment memories of childhood and more from the now more precarious and detached position of "unbelonging"[2] and its ever-deepening sense of that most difficult of "dual citizenships"—insider and outsider. In the six years since I had returned to the province, there had been regular, even daily, doses of headlines announcing "a way of life is dying." Yet, nothing could have prepared me for what I experienced in these travels: the terrible juxtaposition created by the dramatic physical beauty of land and sea and the heavy (dis)quiet of depopulation due to massive outward migration, the antiseptic "tourism tidy" of communities, and the gentrification of homes for summer inhabitants whose childhoods and permanent residences were elsewhere. By day, no smells from busy kitchens, no clusters of chatting men at wharves or garages and, most devastating, no packed school buses, no children shouting playfully in fields, or riding bicycles too close to the road. Plywood covered windows—on churches, schools, garages, stores, government offices, fish plants, and homes—provided two-way shields. By night, the dark sky hung low and sorrowful, drawing out the quiet, accenting the mournful. How could this have happened?

INTRODUCTION

There are many dimensions to any answer to a question about ecological devastation. In this particular case of the dehabitation of hundreds of fishing communities and the outward migration of tens of thousands of inhabitants from Newfoundland and Labrador, Canada's most easterly province, following the collapse of the northern cod fishery.[3] There are the terrible specifics of many levels of fisheries mismanagement and mistrust, along with inflated fish quotas and abuse of and insufficient regulation of fishing technology, along with too many examples of political cowardice. Such bureaucratic bungling interfaced with the sweep of globalization, the acceleration of climate change, the callous and shortsighted destruction of fish habitats, and a general disregard for the biosphere. But, here, in this chapter, I want to focus on a less-often considered dimension, a more affective side of the question of ecological devastation, and to lean into issues of place, attachment, and history to shed some light on how it can come to be that a place and a culture so many profess to love—to elevate to and celebrate as "a way of life"—is now considered by most to be in the throes of a cultural and ecological crisis. How might certain notions of cultural identity and particular cultural beliefs and practices—a sense and sensibilities of place—have been implicated in this catastrophe?

While there are numerous competing and complementary versions of social justice education, for the purposes of this chapter, I want to focus here on a notion put forward by Edmund O'Sullivan (2002), in which he links social justice and ecological justice in a particular form of *transformative education*. Within this notion, O'Sullivan argues for what he calls an integrative transformative education that includes three modes of learning: survival education, critical resistance education, and visionary transformative education (O'Sullivan, 2002, p. 4). Of these three modes, I want to focus, in particular, on the survival mode of transformative education wherein learners confront what O'Sullivan (p. 5) calls "the profound cultural pathology" that is at the core of the devastation that prevails in our contemporary world. O'Sullivan argues that the dynamics of denial, despair, and grief are central to this confrontation within the survival mode. He acknowledges that denial is a defense mechanism that prevents a feeling of being overwhelmed, and despair, a common response when one moves beyond denial, itself has the capacity to overwhelm. Both denial and despair immobilize, albeit differently. O'Sullivan writes, "Grieving is a necessary ingredient in the survival mode. The sense of loss at the personal, communal, and planetary level that is part and parcel of the survival mode demands a grieving process at profound levels....Transformative learning in the survival mode is a learning process requiring the ability to deal with denial, despair, and grief" (p. 5). In this sense, survival education is a learning space predicated on an acceptance of loss.

In this chapter, I explore some of what might be entailed in the survival mode of education in O'Sullivan's tripartite transformative learning process. In particular, I examine aspects of attachment, change, and loss, as they relate to moments in my educational autobiography and to my current research in transformative education and reparation. O'Sullivan, himself, does not articulate, in any comprehensive manner, the dimensions of the grieving to which he refers in his survival mode of transformative education. It is my purpose here to explore more fully what might be entailed in cultural grieving, and, in so doing, to articulate a more complex notion of the work of grieving to which O'Sullivan refers. Framed within an intersection of biography and ecology, the chapter addresses evolving issues of cultural change, loss, and reparation as these relate to the broader cultural context of twenty-first-century Newfoundland and Labrador and the ecological and educational challenges it now faces. When considering what it might mean culturally to move from denial through despair to grieving, other related questions arise: What might constitute a renewed and *reparative* sense of place? What might it mean to speak of relation to place as *ethical relation*? What is the place of education in such shifts and changes? In addressing these questions, I propose a notion of transformative learning informed by and framed within notions of loss and melancholia—a reparative education.

Early Lea(r)nings in Cultural Politics

It is the intimate, never the general, which is teacherly...Time must grow thick and merry with incident before thought can begin.

Mary Oliver, *Long Life*

In 1972, during my final year of public high school in Newfoundland, Farley Mowat, a well-known Canadian writer and conservationist, published *A Whale for the Killing*, an impassioned account of the 1967 killing of a trapped eighty-ton whale in waters near Burgeo, a small community on the south coast of the island of Newfoundland. Mowat and his wife, Claire,[4] had moved to Burgeo from mainland Canada in the early 1960s. Their own conservationist, romantic sensibilities—as well as their critical estimation of what was happening to Newfoundland since confederation with Canada—were challenged by the horrific treatment of the trapped whale that Mowat described as having been shot at repeatedly over a period of days and its back ripped open by the propeller of a motorboat prior to its death. Despite his own physical efforts and his pleas to various levels of provincial bureaucrats and to members of the press, he was unable to save the whale, which died several days later of infection from its many wounds. In *A Whale for the Killing*, the whale becomes a symbol not only of the mistreatment of one endangered species but, more generally, of all species abuse and depletion at the hands of humans. For that reason, the now well-known book[5] is also a plea for greater knowledge, compassion, and humaneness in our encounters with the natural world, a theme recurrent in subsequent writings by Mowat. However, in the 1960s in Newfoundland, most people were outraged by the negative attention Mowat's efforts brought the community and the province. Mowat was vilified as an interfering outsider who took liberties with the truth in order to advance his own conservationist agenda and, in 1967, the disillusioned Mowats left Burgeo.

For many Newfoundlanders of that era and since, *A Whale for the Killing* became a lightning rod for issues of representation and identity. Indeed, to this day, Mowat remains a controversial figure, only recently outspoken against the annual seal hunt that takes place in the Gulf of St. Lawrence (Mowat, 2008). In the mid-1980s, while I was conducting an ethnographic study of the cultural politics of reading and book publishing in Newfoundland and Labrador (Kelly, 1993), many readers I interviewed mentioned *A Whale for the Killing* as a topic of debate and, most often, debasement. The following comments, though but two, are indicative of many:

> I don't like books that tend to criticize, in any way, Newfoundlanders or their ways....I don't mind criticism if it's due, here or there, but if you want to criticize a community [as Mowat does in] *A Whale for the Killing*, then you'd better find someone else to read it....I do not believe it. We are not savages....He is telling lies.

> [Mowat] was not very accurate. In *A Whale for the Killing*, he made the people [of Burgeo] look bad. He thought they used the whale for sport, which...they did. But the whale didn't mean very much to them. If you see five hundred whales, what's one?...I don't think the whale had feelings, to them. They were out for a bit of fun. I know it was wrong and you know it was wrong but in that harsh sort of environment, it was a sport to them. It is no worse than a mainlander...coming to Newfoundland to shoot caribou and taking back the horns and letting the meat spoil in the woods....They are considered sportsmen, aren't they?

Such responses, which questioned Mowat's credibility and noted the hypocrisy of such cultural judgment, were consistent with what I had learned myself as a young Newfoundlander growing up amidst heightened attention to cultural politics and regional inequalities—intense fallout from a deeply contested confederation with Canada in 1949—and amidst what later became known as an artistic renaissance of culturally authored challenges to the stories told of us by others. In such a context, it was difficult to question such lessons in cultural loyalty, solidified as they were, in the case of *A Whale for the Killing*, through the public and institutional reiteration of a monolithic interpretation that renounced the messenger, refused the message, and pronounced as indisputable a right to "a way of life" based on a binary notion of humanity and nature, wherein humanity was independent from and superior to nature. Xenophobic accusations about the writer and charges of embellished writing prevented any serious educational dialogue about ecological ethics and responsibility. This classic expression of "fortress identity"—a conservative and self-righteous sense of worth and dignity in the face of criticism, uncertainty and change (Ang, 2000, p. 2)—fed cultural denial about the existence of—and an examination of implication in—any problem and thwarted any chance of significant progressive ecological change.

Only several years later, in the late 1970s, when I was a young school teacher in a sealing community in coastal Labrador, my students and I would watch sealers harvest pelts from our classroom windows as, during English class, we studied *Death on the Ice*, an account of the SS *Newfoundland* sealing ship disaster by Cassie Brown. At this same time, only a few short kilometers away, while international media watched and recorded, Brigit Bardot and members of Greenpeace painted seal coats to protest what they saw as a brutal, inhumane slaughter. In such a geographical context of economic scarcity and struggle, it was difficult for my students and me to engage the basis of such criticisms of the seal hunt. Greenpeace in the late 1970s seemed to be up to the same thing as had Mowat earlier in the decade: a deliberate misrepresentation of a people to further a cause without any regard for the effects of their actions on those who were misrepresented. It was difficult to imagine a space, back then, where either of these events could be examined otherwise. It seemed you had to choose. It seemed that you could not be a loyal Newfoundlander and be an environmentalist, too—and, I was a loyal Newfoundlander. The lines of allegiance were starkly drawn with a more conservative notion of provinciality or regional identity based on history and tradition—associated with insiderness—on one side, and a more progressive, forward-looking ecological conscience—associated with outsiderness—on the other.

I draw on these early and well-known examples that inform my own biography to begin to raise questions about the obstacles to and possibilities for transformative learning of the sort to which O'Sullivan refers. New editions of the dynamics I recall here have been (re)created regularly in the province in the thirty-five years since the publication of *A Whale for the Killing*, seemingly deepening more so than challenging the lines of allegiance drawn between ecology and provincial identity. Debates such as those around species overfishing, the seal hunt, and hydro, mineral, and oil resource development continue the refrains of *foreign* overfishing, *outsider* misrepresentation, *survival* economics, and *traditional* rights. Rarely, alongside such refrains, is there concerted attention given the broader environmental implications of cultural beliefs and practices and our own serious implication in the problems and issues at hand. At the beginning of the twenty-first century, with many parts of the Province of Newfoundland and Labrador in a literal *sea of despair* as, locally, the impact of the closure of the cod fishery deepens yearly, and as, globally, oceans are emptied of species after species,[6] a fuller reckoning of our interrelationship with the planet and its resources is long overdue. Such a reckoning, which may well be the source of our greatest hope of recuperation and redirection, requires that the ideological basis of incidents of environmental and ecological abuse be examined as cultural and systemic rather than pathological and isolated—a problem in which we are deeply implicated and not one from which we are separate or by which we are (only) victimized. Such a reckoning also requires a move beyond a purely economic discourse toward an economic ethic that forefronts environmental sustainability and planetary consciousness. As James

Schaefer notes, ecology and economy share, etymologically, the same root, the Greek *oikos*, meaning "house." He asks, "Must ecology and economy be a house divided? Won't paying heed to the environment hurt the economy? Here is a false choice. The true alternatives are short-term exploitation and long-term prosperity. Sustainability means a focus on enduring wealth, especially for future generations, rather than mere personal and immediate gratification" (Shaefer, 2006, p. A11).

In Shaefer's remarks, economy is drawn back to its radical roots of the Greek *nomos*, or management, for sustainability and longevity. Such a massive discursive and cultural shift—from presentism and exploitation to futurism and sustainability—entails not only a rethinking of education but also a rethinking of many of our loyalties and attachments, through an examination of the constituent nature of the ties that bind us to one another in culture and place. Most profoundly, this shift also necessitates confronting loss, educationally, through an examination of such questions as "What has happened?" "What part have I played in what has happened?" and "How is reparation possible?" What conceptual tools might help us address such questions?

STRUCTURES OF FEELING AND PLACE

> *It is one of the perils of our so-called civilized age that we do not yet acknowledge enough, or cherish enough, this connection between soul and landscape—between our own best possibilities, and the view from our own windows. We need the world as much as it needs us.*
>
> Mary Oliver, *Long Life*

Much has been written about emotive or affective dimensions of place attachment. Some of this literature addresses some aspects of loss and place. Here, I highlight three specific aspects of this literature of loss as it relates to place attachment: nostalgia, solastagia, and melancholia. Each offers particular insights into contemporary affects of place and the challenges presented by O'Sullivan's survival mode of transformative education. I examine each here to discern these insights, to consider the implications of each for a transformative learning, and to speculate on the possibilities each holds for understandings of cultural loss, mourning, and reparation.

Nostalgia

Nostalgia, a word that has its roots in the Greek words, *nostos*, "to return home," and *algos*, "pain and longing," was originally coined in the seventeenth century by a Swiss medical student, Johannes Hofer, to describe the afflictions of merchants working far from their homes. It was later more widely applied to these same afflictions manifested in soldiers of imperial armies. By the late nineteenth century, its use was more generalized and reflected current popular understandings—a longing for an idealized past time. By the twentieth century, the century of more widespread migration than any that preceded it, nostalgia was accepted as a pervasive, albeit largely regressive, even indulgent affect—an unfortunate but, by then, *normalized* aftereffect of progress and change, paradoxically both widely commodified and as widely dismissed. Yet, as many scholars argue, we dismiss nostalgic urges at our peril, refusing the insight such affects may provide, not so much about the past but, rather, the present. Linda Hutcheon (1998) recognizes this potential of nostalgia when she notes that "the power of nostalgia...comes in part from its structural doubling-up of two different times, an inadequate present and an idealized past" (p. 5). She further argues that nostalgia has an element of both "affect and agency, emotion and politics": "Nostalgia is not something you 'perceive' *in* an object; it is what you 'feel' when two different temporal moments, past and present, come together for you and, often, carry considerable emotional weight....It is the element of response—of active participation, both intellectual and affective—that makes for this power" (p. 5). Hutcheon's notion is one of nostalgia as agenic and politically productive: an embryonic critique of the present, encapsulating a muted hope for the future, misunderstood as longing for an idealized past.

James Overton discusses at length the mobilization of nostalgia in various ways to target the migrant Newfoundland subject throughout the twentieth century. He notes that the nostalgia that has haunted the migrants of twentieth-century Newfoundland, many of whom, as today, left home to work, is "not simply a fantasy, for it guides action and is a powerful creative force" (Overton, 1996, p. 125). Furthermore, echoing Hutcheon, he argues that the utopian images conjured through nostalgia are "less a description of Newfoundland than an expression of current alienation and longing for a more satisfying life" (p. 129). Nostalgia, he argues, is "more than a "homing instinct," for it is also, in many ways, a yearning for a secure and stable existence in a more acceptable world" (p. 137). The urge of nostalgia, then, as reiterated by both Hutcheon and Overton, is both affective and relational, the urge of change and betterment, even if, oftentimes and usually, these urges are framed narrowly, conservatively, and attempt, through these urges, to fix not just a notion of the past, but an essence of identity, as well.

As a migrant Newfoundlander, nostalgia's affect was easily—albeit unwittingly—conjured as both a retreat from identity assault and as a psychic comfort zone, a space within which to remember a sense of cultural self-worth and belonging challenged by the new day-to-day of life as an emigrant. In such situations, it is often hard to discern the extent to which assault solidifies identity or, instead, provides the opportunity for a demonstration of the extent to which identity has been previously solidified. As for many Newfoundlanders, being "away" presented vivid opportunities to confront how this particular provincial identity is constructed nationally as "other." Such displacement offers painful lessons about the terms of identity, the constitutive nature of place, and the cruel terms of difference. No social site is constituted outside of the political. The privilege that accompanies being a middle-class professional is an incomplete shelter from "the stories others tell of us." These struggles were of the same ilk as those the townspeople of Burgeo faced, but I was not on home turf. The response was similar, too—but in the way of a private war—and without national media, it galvanized what was already a strong claim to place: a new edition of past efforts, in which *who we are* is defended, without reckoning, at the cost of imagining *who we could become.* Functioning in these ways, nostalgia was purposeful *and* counterproductive.

Displaced and under assault, my desires, my subjectivity, clung to what nostalgia offered, a space in which to remember the value of what it was I had lost or abandoned, to reject others' denigration of it, and, as importantly, to dream its value differently, more lovingly. As in previous instances of cultural assault, alternatives seemed unavailable. The promise of belonging was elsewhere than where I was. If the construction of identity is an effect felt less in the making than in the enactment or the performance, then the lessons—both explicit and implicit—had been well learned. Nostalgia produces, out of loss, not just expressions of sentiment; nostalgia also produces subjects who seek solace, people who, as Overton notes, act in particular ways in the world. But might some subjects, some constituted identities, be more prone to nostalgia? My own nostalgia had an inevitability about it, a sense that I could not—indeed, should not—feel otherwise. It was much later that I learned how to question not just what nostalgia contains, what it constitutes through its lingerings over a "way of life," but also what cultural conditions create a vulnerability to nostalgia. Such questions can begin to clarify a sense not just of what is lost and what is hoped for, but, as well, of the constraints and possibilities that accompany attachment and separation and the parallel urges of loss and hope and the identities framed through them.

Solastalgia

The emotive force of nostalgia lies in loss, in the impossibility of return, not just physically, to a place, but also to a time and, importantly, to an earlier version of ourselves attached to that time and place. But, what happens when the place to which one is attached is lost in a different way, catastrophically changed for those who continue to inhabit it? Based on his research in the drought-devastated Upper Hunter communities of New South Wales, Australia, Glenn Albrecht has expanded the lexicon of place, pain, and belonging, coining a word that captures the interrelationship, specifically, of ecosystem distress and human distress: *solastalgia*. As Arbrecht explains, *solastalgia* has a structural similarity to *nostalgia*, and its original meaning of homesickness embeds within solastalgia a reference to place and home:

> It is the pain experienced when there is recognition that the place where one resides and that one loves is under immediate assault (physical desolation). It is manifest in an attack on one's sense of place, in the erosion of the sense of belonging (identity) to a particular place and a feeling of distress (psychological desolation) about its transformation. It is an intense desire for the place where one is resident to be maintained in a state that continues to give comfort or solace. Solastalgia is not about looking back to some golden past, nor is it about seeking another place as "home." It is the "lived experience" of the loss of the present as manifest in a feeling of dislocation; of being undermined by forces that destroy the potential for solace to be derived from the present. (Albrecht, 2005, p. 45)

As Albrecht notes, the disease that is solastalgia is firmly constituted around loss. Its physical and mental symptoms are commonly manifested among aboriginal peoples who have been historically and systematically dislocated or whose relationship to the land has been disrupted by development. The applicability of solastalgia to many in Newfoundland and Labrador is obvious when one considers the aboriginal peoples of the province. Elizabeth Penashue, an elder of the Innu, a once-nomadic people of Labrador, has worked for decades to call attention to the distress that settlement, displacement, and environmental degradation have created for her people. An environmental activist and Innu spokesperson, she describes how hydro and mineral development and low-level flying have reshaped the Innu relationship to the land (Ellwood, 1996). The resulting symptoms of this human distress of dislocation and habitat destruction—depression; alcohol and drug abuse; high rates of suicide, diabetes, and heart disease; and the breakdown of family and community culture—coupled with ecological distress evidenced in wildlife devastation and habitat destruction through, among other things, the toxic contamination of waterways and soils, are embraced in Albrecht's notion of solastalgia. The circumstances of the Innu shore up the solastagic effects of imposed assimilation on the cultures and habitats of aboriginal peoples.

Solastalgia is also evident in the devastation and displacement caused by the collapse of the northern cod fishery. Many of the tiny outport communities out of which this centuries-old fishery operated hold names that suggest the degree of solace provided their inhabitants: Little Heart's Ease, Happy Valley, Comfort Cove, Heart's Delight, Paradise, Happy Adventure, and Heart's Content. No longer able to find solace in adequate sustenance from the oceans on which they have historically relied, fishers and their communities have been rocked by the vicissitudes of such loss. In this case of the fishers of northern cod in Newfoundland and Labrador—whether or not it is acknowledged—solastagia is aggravated by direct implication (Albrecht, 2005, p. 50), through fishing practices, in the ecological destruction. This insight is in keeping with the contention of O'Sullivan (2002) of what happens in a shift from refusal, as a form of denial, to the immobilization or inaction that is despair, whereby knowledge of one's implication aggravates, overwhelms, and petrifies. The social and cultural devastation that marks the distress that is solastalgia is well evidenced and well documented.[7]

Solastagia registers a degree of loss very different from that of nostalgia. When a moratorium was declared on the northern cod fishery in 1992, I was living outside the Province. I remember as I stood in a ferry terminal in North Sydney, Nova Scotia, waiting to make the crossing to Port-aux-Basques, I was riveted to a large poster. Despite its depiction of a single dory on tranquil waters surrounded by soft hills, it was not your typical tourism poster designed to allure. This poster, with its four simple words—"And No Fish Swam..."—shocked, saddened, forewarned, and, in its solemnity, suggested the new reality at which we were about to dock. But, upon closer inspection, the fine print just below the powerful message, which read, "A call for action to stop foreign overfishing," revealed another layer of politics, one couched in denial and blame and lacking acknowledgment of one's own implication, as well.[8] A second look, then, revealed a more insidious message, one that implied that "others" were threatening the tranquil and unspoiled picture—the (otherwise) pristine and innocent culture that lay before us.

Distance buffers, and so can privilege, but the ensuing years would soon reveal the effects of the deep psychic crack the ecological devastation that caused the closure of the cod fishery had left in its wake: massive outward migration[9]; family fragmentation; erosion of community infrastructure; closure of schools, churches, and entire communities; and increased social dysfunction and despair. No family has been left untouched. And as the desperation that marks such catastrophic loss becomes opportunity for more monied others, access to the land as a source of solace is increasingly diminished through increased privatization, exploration, and development. Commenting on these developments, activist Greg Malone (2007) writes of the implication to which I refer:

> Our own children on the road to Alberta are themselves the casualties of poor collective decisions. Their parents' generation (mine) was too greedy, we took too much for ourselves and left not enough for them....As ponds, rivers, ocean frontage and lands are privatized, when indigenous and native peoples are cleared off their ancestral lands for commercial development and mega projects, then families, societies and cultures are devastated, and their capacity for happiness is destroyed along with the ancient balance of human life with the environment.... In our anthem we ask God to guard Newfoundland, to protect it. From what, from whom? we used to joke. It turns out to be—from ourselves. (Malone, 2007, pp. 46–47)

The emergence of solastalgia marks a fundamental and even traumatic rupture for the subject—a broken ecological relationship as a tragic reminder of our oneness with nature and of the crucial interdependence that is at the heart of this oneness. Solastalgia, in this sense, denotes the emergence of a planetary consciousness born of a type of catastrophic loss that strikes at the very core of relationship to place. Where human implication and exploitation are factors in solastalgia, as they are in the two examples here from the context of Newfoundland and Labrador, and as they are in so many other examples worldwide, solastalgia bears witness to a historical relationship to place that has been, in many ways, abusive, and, in so doing, it heightens the need for a relationship to place as ethical relation. Whales, seals, cod, people—it is harder, now, to avoid the interconnections—and the errors of our attitudes, our ways. It is harder, now, to refuse to acknowledge the deep cracks in the promises of belonging. It is harder, now, to find solace in identity.

Melancholia

Solastalgia, despite the devastation it marks, is also deeply hopeful in the intense desire for comfort from place that is at its emotive center. While this desire can be the agenic impulse out of which change can begin, solastalgia does not offer a means by which to reconcile the loss out of which it arises, and without such reconciliation, change is thwarted. An examination of the nature of loss itself, its psychic structure and influence, provides insight into loss as it applies to place and to how cultures adapt to and transform themselves in the face of loss. Central to such an investigation is a focus on the psychic dimensions of mourning and melancholia.

As indicated in the introduction to this book, and its earlier chapters, Freud's work is central to any discussion of loss and its vicissitudes. In his work at the beginning of the twentieth century, Freud (1917/1989) initially distinguished between mourning and melancholia, identifying mourning as the natural progression through loss, the process by which normal withdrawal of libidinal energy from the lost object (a person, place, ideal, or object) was completed to allow for healthy attachment to new objects. Melancholia, he argued in this seminal essay, was a pathologized mourning, one in which the mourner could not let go of the lost object and, instead, incorporated the lost object into its own ego. Such melancholic incorporation comes at great expense to the ego, for melancholia not only prevents healthy growth of the ego but it also prevents healthy attachment to new objects. Furthermore, melancholia enhances feelings of ambivalence toward the lost object, which is both loved and hated, feelings that are often turned on oneself and fed by guilt and shame. Importantly, Freud later revised his thinking on melancholia, arguing that melancholia was less a pathology and more a step in the process of full and healthy mourning.

Eng and Kazanjian (2003), in their overview of the long history of melancholia, from antiquity to the present, from humoric theory to psychoanalytic theory, conclude that, given the historical legacies of trauma and loss of the twentieth century—the century that has seen more violent upheaval, death, and displacement than any others—that "melancholia at the turn of this [twenty-first] century has emerged as a crucial touchstone for social and subjective formations (p.23). Furthermore, they note that "[a]s both a formal relation and a structure of feeling, a mechanism of disavowal and a constellation of affect, melancholia offers a capaciousness of meaning in relation to losses encompassing the individual and the collective, the spiritual and the material, the psychic and the social, the aesthetic and the political" (p. 3).

Calling melancholia, after Freud, "a confrontation with loss through an adamant refusal of closure," Eng and Kazanjian argue for a notion of melancholia that is hopeful and agenic, one in which the disavowal of loss that is melancholia signals less a fixing or freezing of a past and more an ongoing active relationship and engagement with the past. They further argue that melancholia, understood this way, is productive both as a necessary part of successful mourning and as the basis from which hopeful, healthier futures might be envisioned.

Following Eng and Kazanjian, a melancholic culture, then, can be understood as one that has not only suffered innumerable losses but is one that is (still) *actively grappling*—through obsession and repetition and manifested in the symptoms of nostalgia and solastalgia—with those losses. It could be argued that Newfoundland and Labrador, with seventy percent of its population of Irish ancestry, has inherited some of what scholars have identified as the "melancholy fatalism" (Lloyd, 2003, p. 221) that has been a historical and inhibiting feature of much of Irish culture. In my own initial research on Newfoundland culture (Kelly, 2003), I noted how the legacy of this melancholy registers, still, as both affect and subject position, but, despite some of the commonalities of colonialism and imperialism, which Ireland and Newfoundland and Labrador share, it does so with its own persistence, particularity, and twists. Not the least of these differences is the tension between the regret of lost nationhood and the felt subjection of its place within the Canadian Confederation[10] and global economics, the circumstances of which continue to fuel and to deepen identity politics.[11] And in many ways the identities defended are often essentialized and conservative leaning, ones in which, as Ien Ang (2000) notes, "identity is firmly conjoined with the very antithesis of change" (p. 6). Yet, such a reading might oversimplify and ignore the complexity of the historical subject, one who, while entrenched and, perhaps, disoriented, is yet passionately negotiating the shifting cultural and political ground beneath its feet? John Dalton suggests that "every melancholic confuses the unknowability of the future with the apparent nothingness of the present, a nothing that can be thought otherwise as a 'not-yet,' the space of potentiality. Our mourning, then, is perhaps not first of all the recursive experience of loss, but a symptom of being overwhelmed by a responsibility: for one's own finitude, and for the acknowledgment that finitude is the condition of the aspiration for change. Nothing is stable" (Brown et al., 2006).

If no fish come, what becomes of a place the raison d'être of which was fishing? So many, in economic desperation, sing the refrain *There is nothing here for me, now. I've got to leave. What else is there for me to do?* This refrain, in all its variations, can be read as much as a statement of fact about loss, its implications, and its impossibilities as an expression of desire for something different—the "not yet"—a longing overwhelmed by the implications of finitude.

If melancholia can be understood as a form of attachment through which is expressed a longing to build on what is lost, to create new understandings as the basis for moving forward thoughtfully and reflectively, then, melancholia is an affect tinged with, as well as frustrated by, the seeming unattainability of the goal of justice, the urge to do right by that which has been lost, or gone: *to repair*. In melancholia, those who have suffered loss are not yet done with the lost object; something must be righted, repaired. Melancholia, then, while incorporating the agenic instincts of nostalgia and the manifested distress of solastalgia, has a more extensive, productive, and hopeful reach. The basis of this hope is reparation itself. Melancholia provides insight into loss and into the structures of feeling that can disable and inhibit change. But, in reparation, it also gestures to the ethical basis for hope. As numerous scholars (Eng & Kazanjian, 2003; Cheng, 2003; Butler, 2004; Brown et al., 2006) now argue, such attachment as is seen in melancholia warrants rereading and reconfiguration as a rich conceptual and political tool of contemporary struggle, one that has the potential to highlight cultural affect, reparation and healing.

Both nostalgia and solastagia reference melancholy as it is used in the popular lexicon, to denote an affect, a sadness or pensiveness. While premised on loss and an understanding of loss as having both emotive and physical registers, accounts of neither nostalgia nor solastalgia explicitly articulate the conceptual basis of loss itself. Nor does either adequately question the constituent nature of the attachments out of which these structures of feeling arise. Melancholia, however, provides explicit conceptual connection to and insight into the processes of loss and mourning. In this sense, melancholia offers a basis from which to address the possibilities of and for transformative learning and cultural change while also expanding the basis from which the nostalgic and solastagic subject may be understood and transformed. What might it mean to move beyond the cultural habits of denial and disavowal toward acknowledging loss—mourning what has been lost and taking responsibility for the ways in which we may be implicated in what is lost—and creating new practices of reparation, healing, and renewal? How might education help in this project of confronting loss and our implication in loss?

LATER LIFE LESSONS: (RE)INVENTING ECOLOGICAL SELVES

In childhood, only the surroundings show, and nothing is explained. Children do not possess a social analysis of what is happening to them, or around them, so the landscape and the pictures it presents, have to remain a background, taking on meaning later, from different circumstances.

C. Steedman, *Landscape for a Good Woman*

As we direct children's gazes, we introduce them to the world we care about, and so their notice of the world and ultimately the world they see is the one we care to bring to their attention... children cannot get the world without someone who points to a world worth seeing.

M. Grumet, *Bitter Milk: Women and Teaching*

Caught in the vastness of natural abundance, a child would be forgiven for failing to comprehend fully nature's vulnerability, its temporality. A child would also be forgiven for learning too well the cultural lessons taught it, even when these lessons are terribly limiting or simply wrong.

As a child, I learned well to love my place, to bask in its dramatic beauty, to find adventure in its woods, to explore mystery in its low night skies, to seek solace by its rivers and ocean, to play with abandon in its hills and fields, and to feed off its plenty. This loving relationship was always deeply spiritual, even when such words were not available to me to describe it. And while I did not learn to love without also learning that love entails responsibility—to not unnecessarily harm, to not pollute or litter, to not jeopardize—alongside these loving lessons, as a child, I also learned other harsher ones—about cultural violence and abuse—that (continue to) indelibly mark me. Examples such as the senseless torment, torture, and unnecessary poaching and killing, by some, of "big game" animals for fun or "sport," the reckless disregard, by some, for spawning fish and their habitats for short-term economic gain, and the arbitrary violence, of some, that could be directed at domestic pets, while not confined or peculiar to this place are, nonetheless, a real part of its particularity. Existing side by side with loving practices, they form a layer of connective tissue that links micro and macro levels and understandings of experience. I did not learn, until I was much older, the larger social, cultural, and ecological concerns to which these vicissitudes connect. To excuse these behaviors as part of a "character" of a place—any place—is to refuse to consider the wide-ranging implications of the belief systems that underpin and "rationalize" them—in whatever place they happen to occur.

Jonathan Lear claims that the "inability to conceive of its own breakdown will tend to be the blind spot of any culture....A culture does not tend to train the young to endure its own breakdown...This is not an impossible thought to teach, but it is a relatively new idea in the history of cultures...A culture tends to propagate itself, and it will do that by instilling its own sense of possibilities in the young" (Lear, 2006, pp. 83–84).

By way of a culture propagating itself, as a child, alongside other cultural lessons, I also learned about identity and the conditions of belonging, a lesson in love and attachment (which is always also a lesson in abjection and hatred)—a lesson that, in this case, unintentionally circumvented my fuller ecological growth. I mistakenly learned to take for granted my claim to this place and its natural bounty. I mistakenly learned that, in relation to this place, it was *I* who was vulnerable, at risk. I did not learn until much later that vulnerability is mutual, that this place and I are *interrelational*, interdependent. Most unfortunately, I also learned—and these lessons, an underside of a provincialism, a neonationalism—that tradition can trump conservation and that community solidarity too often trumps constructive criticism.

But these latter lessons were discomfiting, forged ambivalently—evidence of an attachment not fully secured, its loose end an example of "a crack that lets the light in" (Cohen, 1992), the thin edge of a wedge that allows (for hope of a) necessary separation and growth. All of these learnings to which I gesture were not individual but collective; they were not arbitrary but systemic; and, they are as much a feature of "our now" as "my then." These learnings were and are cultural and ideological—part and parcel of what it means to grow up in and of a place "founded in" the destructive logic of gender, race, and class injustices, a place, like so many others, where, often, one is encouraged to love what one should, instead, question, and where one is, too often, discouraged from change in the name of loyalty and tradition. Such lessons are not uncommon in any culture that is steeped in imperialist and conservative ideologies through which resource exploitation and greed are *naturalized*, in the name of profit and power, along with their accompanying deeply regressive logics of development, growth, and progress. The subjugations of globalization—the intertwined exploitations of places and peoples—is but the latest face of these historically embedded discriminations.

Adults owe more to the world—and to its children—than mere cultural reiteration. It is the responsibility of adults to be reflective, to attempt to reach beyond the confines of cultural ideologies and traditional beliefs. It is the responsibility of adults to learn as a pursuit of meaning, as a project of responsibility, of legacy, of ethical living. No sector escapes this responsibility to children but, it might be reasonably argued, educators hold a heightened degree of it. My early and brief harkening to a time in the 1970s when I was a public school teacher is a specter of pedagogical implication: the cultural politics of pedagogy. There was much more to examine, in those early days, than the politics of representation and identity that arose out of the antics of Greenpeace and its efforts to eliminate the Newfoundland and Labrador seal hunt. Loyalty—unthinking and unconditional adherence to the demands of identity—can make fools of us all when we fail to consider the grounds on which it is forged, its often simplistic and harmful battle lines, and what it requests of us, in the name of being loyal subjects. Back then, teaching *Death on the Ice* as Greenpeace protested on ice floes only a short distance from our school, I thought I had grabbed a teachable moment when, in retrospect, I had missed an opportunity. I got it only partly right, unwittingly compromising, through my own particular disavowals, what I now see was the educational responsibility to question the grounds of identity, to consider the implications of all our cultural practices, to study what we might *become* as much as what we are and were, and to imagine—through forthright dialogue—how our places, our world, need to be different. Even now, I do not endorse such campaigns as those that continue to be waged to a media frenzy on the ice floes off Newfoundland and Labrador each spring.[12] But I question, along with others (Frampton, 2008), the cultural self-righteousness, in its many guises, that can too often be the standard response to any criticisms of the seal hunt. Teasing out the complex layers of contradiction that sediment our everyday meanings is part of the hard work of reconciling ourselves to implication and a necessary first step in reparation.

A reparative, transformative teaching would begin from a different set of assumptions than those out of which my childhood learnings were derived and on which my initial teacher preparation program was founded—ideology presented as a set of pregivens instead of a set of (always ever questionable) assumptions and practices—and those I attempted to enact in that school classroom so many years ago. Those assumptions, driven by empiricist and rationalist ideologies, are infused by a notion of self as predetermined rather than made, and self and others as separate from and superior to the natural world. Such assumptions attempt to maintain and solidify existing beliefs and traditions and, when enacted in education, they stifle debate, creativity, and change: education then becomes a dinosaur-maker. Davis, Sumara, and Luce-Kapler (2000) propose, instead, an *ecological postmodern* perspective on identity and teaching, one that acknowledges the profundity of identity, self as emerging from and constituted by biology, environment, culture, history and the cosmos, a self that is "both a product of complex processes and a complex process that participates in its own making" (Davis et al., 2000, p. 178)—a self that can be undone, remade, a self that can change and participate in its own change and that of others. Teaching, understood from this perspective is "all about effecting transformations" (Davis et al., p. 178), and the degree of implication—the implications of the smallest educational gestures and the implication of education in larger patterns—is staggering:

> In encouraging particular sorts of understandings, the teacher is supporting the development of particular worldviews and modes of perception. The associated classroom experiences are biological-and-social events. On the sub-personal level, for example, they contribute to actual physical transformations in brain structure, as well as to other physiological changes. On the personal level, they frame how one sees and acts. On the interpersonal level, they influence collective dynamics as they affect how people think about and relate to one another. On suprapersonal levels, they are enfolded in social and cultural patterns, which in turn impact on the grander systems. (Davis et al., 2000, p. 178)

Such aspects of the remaking of self, perspective, and place are echoed in many calls for an ecologically sensitive education. David Gruenewald focuses on two objectives of what he calls "a critical pedagogy of place" (2003, p. 9): decolonization, which resonates with the deconstructive base of O'Sullivan's critical resistance education, and reinhabitation, which resonates with the creative base of O'Sullivan's visionary transformative education. For Gruenewald, decolonization is a means of becoming conscious of loss that "involves learning to recognize disruption and injury and to address their causes...[and] unlearning much of what dominant culture and schooling teaches" (p. 9). Its complement, reinhabitation, involves "learning more socially just and ecologically sustainable ways of being in the world" (p. 9). Decolonization and reinhabitation are concrete educational and ecological goals of a culture marked by the effects and affects of environmental distress and loss, a *melancholic* culture looking to do right by that which it has not yet fully acknowledged is lost.

These objectives proposed both by O'Sullivan and Gruenewald are fundamentally reparative in nature. In a world of catastrophic injury on so many levels, perhaps it is time to begin to speak more explicitly and clearly about a *reparative education*, one that encompasses a project of confronting loss, addressing melancholia, and promoting healing and renewal. Reparation, for example in the sense promoted by Melanie Klein (1975) and others, is closely related, conceptually and psychically, to notions of melancholia and mourning and is, as well, already an established metaphor in relation to justice and some of its best inclinations. Melancholia suggests an implication in what is lost, and the reparative urge is the urge to right that wrong in which one is implicated. As an ethical urge, then, reparation—reparative education—may well be a very timely and compelling metaphor for our times—and for education for our times. As John Cairns, Jr. (2003), notes, in a discussion of environmental reparations, "Repair of ecological damage is an act of enlightened self-interest, as well as an ethical imperative....The long-term hope for the human species—sustainability—is a constructive, compassionate approach. Regardless of what happens to humankind, it is probable that some species will survive until the sun fails. Even if the human species does not, it seems ethical to make an exit that is notable for acts of compassion rather than acts of rage and revenge" (p. 25).

Conclusion: Reparative Attachments

What is allowed to live when something else dies? What is opened as a possibility when something that has claimed us is finally put to rest?

Wendy Brown

A Short History of Progress by Ronald Wright (2004) is an influential and urgent call for ecological change. In it, Wright draws on a well-known analysis by Joseph Tainter, in which he outlines three, usually interrelated, features of—descriptive metaphors for—cultural collapse: the runaway train; the dinosaur; and the house of cards. Upon closer examination of these metaphors, each is evident in the cultural crisis occurring in Newfoundland and Labrador. There is *the runaway train* of technology-based overfishing, resources depletion (fish stocks, wildlife), new resource exploitation (oil and gas), economics-driven outward migration, strained social institutions, increased social stratification (between richer and poorer, urban and rural) and increased social dysfunction (violence, abuse, drug-related crime, addictions, suicide). There is *the dinosaur* of resistance to change, denial of seriousness of situation, blame and abjection, collective inertia, a fierce clinging to tradition and custom, and an inability to imagine and to proceed differently. Finally, there is *the house of cards*, in which one card leans on and affects what happens to all the others, and that captures the fragility of an interrelated, insufficiently diversified, complex, and, ultimately, presently unsustainable web of life. These features are the sum total of our desires, devastations, and denials.[13]

O'Sullivan (2002) notes, "any examination of quality of life must attend to our deep-seated need for community and sense of place" (p. 9). Deep connection to place is a necessary precursor to passionate caring for a place. A strong sense of community is essential to effective action. In Newfoundland and Labrador, both connection and community still exist in abundance and are key strengths from and with which to move forward. But attachment to place is not inherently good or bad. Many forms of attachment are inhibiting and debilitating; all forms of attachment must be scrutinized and troubled. A deep sense of place must be a deeply ethical sense of place, or place itself—in its deepest and most profound sense—cannot be protected, maintained and healed. To this end, attachments cannot be acritically celebrated; forms of attachment must be examined for what they tell us about ourselves, as subjects of a place—and, upon scrutiny, it is entirely possible that we may not like what such scrutiny reveals. But, such examination can lend itself to necessary separation and can lay the groundwork for detachment, reattachment, and reinhabitation.

Rebecca Martusewicz (1997) reminds us, "Detachment is at the heart of education and, thus, of our ability to think of a better world" (p. 15). She writes, "Detachment is a matter of shifting away from certain homeplaces. Psychological, political, or geographical, these places provide a sense of comfort, and thus, separation necessitates a kind of grieving. This grieving is a result of the loss of comfort, security, and pleasure that gets made relationally in homeplaces—those people, beliefs, rituals, and events that touch us, that move us, though they may paradoxically be unhealthy for us" (Martusewicz, 1997, p. 16).

It is such detachment that locates one, simultaneously, as both cultural insider and outsider, as a subject of place constituted in loss—a loss with potential to enable change. Melancholia, as a response to loss, is understandable and, arguably, a necessary and potentially fruitful feature of attempting to honor what is lost. However, habitual, unmediated, and unquestioned melancholia, or incessant disavowal, inhibits the forward movement for which its very presence suggests a muted longing. For this reason, concerted public spaces in which to dialogue about the implications and effects of loss and to represent and to ritualize loss are central to confronting it and to realizing the cultural hope that arises from it. Wendy Brown recognizes the profound shift that is the basis of such a project, what she describes as "a tectonic shift in political subjectivity, identity and orientation....And to arrive at that shift, outside of a melancholic or aggressive frame, mourning is necessary. Aggression is what emerges in the spaces of unmourned losses" (Brown et al., 2006, p. 31). The appraisal of loss and its attachments is a project of transformation that involves both undoing our cultural selves and remaking those selves in light of what loss teaches us: moving through melancholia to mourning. Judith Butler (2004) suggests, "One mourns when one accepts that by the loss one undergoes one will be changed, possibly for ever. Perhaps mourning has to do with agreeing to undergo a transformation...the full result of which one cannot know in advance. There is losing, as we know, but there is also the transformative effect of loss, and this latter cannot be charted or planned" (Butler, 2004, p. 20). In this sense, melancholia is always tinged with fear as a form of the refusal of the risks that are inherent in mourning.

Part of the project of a reparative education is to facilitate the acknowledgment, in all these dimensions, of loss itself, of what is lost and what is threatened to be lost, and our implication in loss. In so doing, reparative education propels forward a sense of cultural responsibility, encourages detachment as a practice of courage and justice, and forges more ethical, reparative attachments as a practice of renewal and hope. In this sense, the cultural mythology of survival that has been generated throughout the history of a place will also be renewed. This work is the challenge implicit in O'Sullivan's sparsely defined survival mode of transformative education. It is the *explicit* and defining work of a reparative education, one informed by a deep knowledge that grief is not the end of love but, rather, the place from which we begin to love anew. The difficult losses that are part of the colonial and confederate history of Newfoundland and Labrador have wreaked great cultural pain; the work of reparation, while hopeful, will undoubtedly be as difficult as it is necessary.

Where Hope Resides...

When I returned to Newfoundland, I encountered what I came to recognize—in myself and in my home province—as a shroud of melancholy, woven through historical disposition and worn as cultural habit: a condition of identity. The thick texture of loss was everywhere—in the music, through the verbosity, beneath the humor, behind the smiles, despite the bravado. Reconciling myself to living within this ethos of loss and finding my own critical place within it, in relation, has become a focus of my teaching and scholarship. Until now, here in this chapter, I had resisted writing about "my place" in relation to environmental politics and ecological concerns. My reticence—tangled thread of loyalty that it was—implicated me further in the sorrows at hand. Beginning to confront some of the complex contradictions into which I am inserted is reaping difficult yet important insights. These lessons are manifested, now, through a teaching and writing as reparative gesture—to rekindle hope, to catch anew a glimpse of a dream, and to learn a different way to belong.

Postscriptum: "Learning to Love Again"

It has been a somber evening in my graduate class on teaching and learning. The theme of the three-hour discussion—identities in relation to place, what they offer, what they inhibit—is close to everyone in the room. No one is untouched by these hard matters of belonging. There has been talk of the massive cultural shift that many feel permeates everything here—including who they think and dream they are—in these times since the cod fishery collapsed. So much seems threatened, or lost completely. Uncertainty prevails, as does a not knowing what to do—the hardest feelings with which to live. Emotions run high in the room, sorrow running a close second to anger, the chief among them. Toward the end of the class, a quiet seems to settle in, a fatigue more so than a pause. One student enters it, his comment offering what appears to be the evening's epithet: "Whether we stay or leave, there is anxiety and grief. If we leave, there is the loss of connection as a result of distance and displacement. If we stay, we face other harsh realities: an aging population, death of outports, and, with oil development and urbanization, bigger gaps between rural and urban and richer and poorer. Despite our show of provincial pride, our relationship to Newfoundland and Labrador is a love-hate one. We love to love it, to show it off to the world. But we hate the depression, the gloom we often live in, and the new rite of passage at ship and airport terminals around the province." A sea of silence follows these comments. These are teachers. They know well how necessary hope is to growth and survival: as necessary as love. And this comment does not sound hopeful. But they are beginning to recognize—and want to resist—the profound tug of melancholy. Perhaps this is why this comment cannot and will not be the last word for the evening. Another voice enters: "Listen here," she says, her barely audible voice cut through with a mix of pride, pain and resolve. Turning to the last page of one of the assigned readings for the evening's class, Greg Malone's short essay, "On Happiness," she reads, "In any ordinary life of inevitable suffering and loss, it is hardly possible to escape those crucial moments where we must decide how we will respond to adversity. Choosing to love, to dedicate oneself to the creative care of our people and planet, is our best option" (Malone, 2007, p. 47). She looks up. "We've all read this. At some level, we all know this. Now we have to learn to live it. We have to *teach* to live it. When he was here, David Suzuki called on us to ask ourselves how much we love this place, and let that love be our guide to ecological action. It's time we did that. It's time we loved this place differently." A better ending; the best possible beginning...

Chapter 5

The Place of Reparation
Loss, Ambivalence, and Teaching[1]

In the Fall of 2000, when Jim Garrison and Dan Liston invited me to contribute a chapter to an edited collection they were compiling on the place of love in teaching and learning, I had just returned to Newfoundland and Labrador, my place of birth, to begin a position at my alma mater, Memorial University of Newfoundland. As I have written in Chapter 1 of this book, that return was prompted and marked by much chaos and ambivalence. I attempted to write my way through some of these difficulties in the chapter I wrote in response to their invitation. I began with the questions, What brings me to teaching? What do I bring to teaching? Out of what circumstances and struggles do these questions present themselves? I set out to explore these questions in a manner that might help me confront the disquiet that has haunted my relationship to teaching since the death of my mother two years prior to my return home. This chapter, a revised version of the one written for and published in that edited collection, entitled *Teaching, Loving and Learning*, is an attempt to bring a notion of reflective grief to a study of my teaching and to speculate on the place of such inquiry in an educational discourse of loss.

Beginning Places

In her psychoanalytic inquiry into teaching and learning, Deborah Britzman focuses on the intricacies, difficulties, and possibilities of a theory of education that "can tolerate the vicissitudes of love and hate in learning [and] that can begin with a generous curiosity toward the subject's passionate capacity to attach to the world" (Britzman, 1998, p. 20). Reiterated in her prolonged argument for a theory of education that centers education as a psychic event is the importance of reckoning not just with the struggles of the student with and in learning but also with those of the teacher. At stake is how we understand what can be accomplished through education—or what education should attempt to accomplish—and in what ways, as the intersection of the psychic and the social construct parameters for any educational project. In this respect, teacher reflection and self-analysis are the means by which one answers the ethical challenge to grapple with one's investments in, passions for, and refusals of teaching and learning. Yet, this challenge comes undirected, even if not directionless. As Britzman (1998) acknowledges, "self-analysis must be subjective and oddly singular, just like the vicissitudes of learning. There are no directions to follow, no grand plan to ensure consistency, no guarantees" (pp. 47–48). And the project is an interminable, ever-confounding one, a continual confronting of the surprise moves that are the emotional work of teaching and learning.

Several rich examples of the kind of self-reflective work to which Britzman alludes already exist (Boldt, 2006; Salvio, 1998; Todd, 1997; Gallop, 1998; Simon, 1993). In each of these examples of self-analysis, the authors wrestle with the complex and contradictory work of looking while seeing (only) partially so as to find yet another place from which to begin to look anew. In such a process, insight can be quickly transformed into another blind spot in the perplexing interplay of experience and reflection as they unfold. Judith Butler (2008) notes, "My account of myself is partial, haunted by that for which I can devise no definitive story. I cannot fully explain why I have emerged in this way, and my efforts at narrative reconstruction are always undergoing revision. There is that in me and of me for which

I can give no account" (p. 37). Still, it remains important to pursue insight as partial, contingent, and recursive unfoldings, as part of the potential of a new ethics of relationality (Butler, 2008).

One obvious place to begin an investigation, however limited, of one's investments in teaching and learning is with the immediate struggles that might suggest a haunting pattern of stubborn preoccupations and ongoing contradictions. Thus, in an effort to begin to articulate the basis of some of my own educational attachments, this chapter continues the investigation of the seemingly banal nexus of biography, geography, and pedagogy as psychic passions, in this case, as constituted (and constitutive) sites of ambivalence in teaching and learning. Specifically, I strive to establish a series of connections among telltale passions with a beginning investigation of what might lie beneath a lifelong set of loyalties, beliefs, and practices.

In pursuing this series of connections, I focus more explicitly here than elsewhere in the book on the psychoanalytic notion of reparation as developed by Melanie Klein, and its usefulness in an analysis of autobiography and teaching—and teaching, generally. My main purpose is to unsettle taken-for-granted expressions of a desire to teach by asking, in my case, on what impulses might such a desire be based and what (reparative) urges might constitute my own love of teaching? Secondly, I explore the relationship of love, a love of place, and a love of teaching to reparation and teaching. Finally, I discuss a notion of teaching as reparative as I address some aspects of a pedagogy of reparation. In so doing, I focus explicitly on reparation and teaching within a teaching rendered overtly political, the desires of which reach toward justice and education across social and cultural difference. While each of these discussions is decidedly autobiographical, each also posits a series of theoretical questions that reach outward, beyond the singular, toward answers that reside, invariably, in the autobiographies of others. Autobiography is, after all, not how we reveal the world but how we forge connections to and meanings from our encounters in it.

Making Reparation

Every story is a story about death. But perhaps, if we are lucky, our story about death is also a story about love.

Helen Humphreys, *The Lost Garden*

No words mean as much as a life.

Anne Michaels, *The Weight of Oranges / Miner's Pond*

I began to revisit the work of Melanie Klein while investigating the relationship of grief, mourning, and identity, a pursuit impelled by emotional chaos following the death of my mother. This chaos is articulated by Joan Didion (2005) in *The Year of Magical Thinking*. She writes about not trusting herself "to present a coherent face to the world" (p. 168) and of having to revise her understanding of what she had seen in the faces of others who have suffered loss. "What struck me in each instance was how exposed they seemed, how raw. How fragile I understand now. How unstable" (p. 169). This instability registered on many sites and, as profoundly disruptive, on the site of professional identity, that is, around a particular notion of teaching and my self or selves as teacher. The profundity of grief is well known, yet its effects feel always oddly singular—as are its manner(s) of disruption. Didion (2005) refers to "grief as we imagine it and grief as it is" (p. 189), noting that in "grief as we imagine it," there is no sense of "the void, the very opposite of meaning, the relentless succession of moments during which we will confront the experience of meaninglessness itself" (p. 189) that accompanies "grief as it is." Yet, grief's void, its challenge to our established meanings, is also a profound opportunity. For example, Richard Johnson writes of specific forms of grief's disruptions as they relate to challenges to regressive meanings of sexual identities, arguing how, in his case, in the process of grieving the loss of his wife his "investments in heterosexuality and certain kinds of masculinity" (p. 16) were destablized. He argues that mourning, as a reflective process, is productive in that it can provide the basis for the reconstitution of identities and a renewed understanding of one's life work. In what Johnson calls these "grievous recognitions" (Johnson, 1997) that accompany reflective mourning are new questions, new directions, and renewed commitments to one's life work.

For Johnson, the gender dynamics that constituted a point of conflicted attachment for his wife, Jill, and him, became the source of reflection and change following her death. More generally, loss provides an opportunity to revisit our attachments and their often conflicted, ambivalent characters. Such recognitions, as was the case for Johnson, can be *radically reparative* when constituted by a particular politics wherein the urge toward reparation is part of what it means to reach and to educate toward social change and justice. Just as Johnson saw himself "doing justice" to his wife following her death, so too do many of us attempt retroactively to repair harm. Such a notion of reparation and its relationship to love and loss form the basis of an examination of some of my own choices in teaching and constitute grievous insights that renew and redirect.

Revisiting Klein's work, following Johnson, provided an opportunity to engage her explanations of key aspects of the issues with which I struggle. Klein (1975) contends that any situation of mourning reactivates early psychic or developmental processes, in particular, what she calls the infantile depressive position, coincident with weaning, in which the child mourns the loss of loved objects (the mother's breast) and all that these objects have come to symbolize (love, comfort, security, and satisfaction). According to Klein, at this stage in its development, in its unconscious mind, the child develops feelings of concern for the loved object borne out of feeling responsible for its harm or loss because of persecutory feelings or fantasies of aggression and hatred toward it when the loved object was not available or able to satisfy the needs of the child. In this depressive position, love and hate coexist, and ambivalence develops as a safeguard against hate. As the child contends with its own felt aggression and these destructive impulses toward the loved object, guilt ensues. From these feelings of guilt comes the desire *to make reparation*, to repair, to make good on the injuries done to the loved objects.

Through reparation, largely accomplished through having and creating experiences of love in the world, the (internal) loved object is restored and a foundation for security, trust, a belief in goodness, and love in the external world is established. With the internal and external worlds so interconnected, the ability *to make reparation* becomes an essential quality in the health of both worlds, as it is in making reparation that integration is accomplished. Eve Kosofsky Sedgwick (2003) sees the depressive position as "anxiety mitigating" and one from which it is possible in turn to use one's own resources to assemble or "repair" (p. 128). In this explanation, reparation is recognized as an essential character of all expressions of love as the conflicts out of which it arises, and its expressions, are transferred to new relationships as the child's world expands and grows. That is, residues of these early dramas accompany us throughout our lives as part of the struggle to contend with the conflicts defined in the early stages of development and to secure for ourselves a loving place in the world.

Klein argued that, when mourning an *actual* loss of a loved one in death, the need for reparation is intensified as the psychic world of the mourner is imperiled, thrown back, as it were, to these early stages of development and the good object, lost in death, is, once again, jeopardized by the bad, and thereby the threat of dominance by the bad object is renewed. In this manner, and as noted in the discussion of O'Toole's memoir in Chapter 2, one grief can provide what Johnson calls "a more cathartic basis" (Johnson, 1999, p. 17) for other, earlier griefs. When new losses are mourned, ambivalence and mistrust are once again heightened as the psyche struggles to contend with the guilt aroused by the reliving of fears and anxieties associated with the loss of the loved object. In this state, relationships (with friends, partners, colleagues, oneself, etc.) are questioned as the ability to trust and to feel understood is jeopardized. Normal mourning, Klein argues, is the process of reparation, through which the inner world is repaired and rebuilt and a belief and trust in the existence of good objects is reinstated and a sense of peace is achieved. In these ways of (living in) love and loss, then, reparation may be understood as (re)constitutive, productive and structuring. Herein, loss not only connects us to other, earlier, losses; it also connects us to others, for loss is our common human denominator. In her discussion of loss and mourning as the basis of a new politics, Judith Butler (2004) notes, "Despite our differences in location and history, my guess is that it is possible to appeal to a 'we,' for all of us have some notion of what it is to have lost somebody. Loss has made a tenuous 'we' of us all" (p. 20). If we examine the constitutive basis of our individual losses, we can reap a deeper connection to an always vulnerable collective—a goal for education across social and cultural difference.

My relationship with my mother had always been deeply and consciously implicated in my teaching. Her teachings were and are an active and visible presence in my work—a busy, ongoing refusal to forget her before she was ever gone. Our connection was deepened by the fact that teaching was something we had in common while mothering was not. Following her death, the imperilment described by Klein was most deeply felt through a deep disorientation and dishevelment around my (teaching) work, a questioning of my confidence and faith in teaching, and a growing distrust of the intentions, possibilities, and effects of teaching for social change. Yet, why was so much of grief's disruptions located within my work? What was I to do with this gnawing, nearly immobilizing loss of faith? As the complexity of this implication of loss in my teaching, with all its nuances and intricacies, continues to unfold as grievous insights, I have begun to rediscover teaching as a practice of love and reparation. In attempting to understand the profound impact of the (life and) death of my mother on my own life work, I began to explore and come to terms with the profundity of grief, its relationship to love, and, for me, the relationship of love and grief to my teaching. Understanding my own reparative urges is a crucial aspect of these realizations and reconciliations.

The work of teaching and learning has preoccupied my life. As I reconstruct my story of teaching and learning as ongoing post mortem, its thematic gist seems deeply reparative. When I was a young child, school achievement was part of the way I secured a place with my mother. She was a busy, overworked mother with a husband away at work and, therefore, raising a family largely alone, so our intimate moments where those carved out to attend to my schoolwork, a mutually purposeful space in which I held her attention and she held on to some vestige of her displaced teaching career. Her decision to leave teaching upon marriage, while in keeping with the expectations of women in those historical times, was nevertheless her interminable regret, solace for which I seem to have sought for her in my own choice of work and study in teaching. Literally, I would go on "to profess" teaching, that which she left in exchange for marriage. In "doing her justice" in this way, I also negotiated that which she had lost—independence and separation—as I simultaneously solidified her and my connection. Achievement is one way we can convince ourselves that we are worthy of love. In my case, my achievement could mean that I was not a burden—as children must sometimes feel they are to overworked parents—but a reward. Attachments to admired teachers were likewise secured through achievement, as I became the "good student." But achievement can garner resentment as well as respect. Like many, I learned to manage any potential resentment I might receive from others through self-effacement and a diminishment of the very achievements on which I relied for approval. Through this enacted split, I learned to never fully realize, accept, feel comfortable in, or celebrate achievement. What I learned instead was to self-deprecate, to regard achievement with great ambivalence, as asset and as liability, as something which secured some love attachments but threatened others.

My struggle for love and approval existed in an educational nest. Within teaching, I—as, in varying ways, do all teachers—unwittingly confront and recreate elements of the most challenging and informing psychic dramas of familial relationships. The structures of power that organize teaching demand that it be a site on which one can be heard and on which one must be attended to. But, it is also a place where one feels and confronts other emotions, such as intense dislike. Teaching is thus a likely place in which are staged psychic dramas of love, hate, and ambivalence—the unconscious return or reenactment of these old dramas (Salvio, 2006; Pitt, Robertson, & Todd, 1998). As well, teaching is an endless rehearsal of separation. At all levels, teaching is structured in strictly defined temporal arrangements. Along with love, the domain of teaching is marked by loss and necessary leavings.

This story (of mine) is not uncommon. That teaching is fraught with the perils of object relations, as elaborated by Klein, has been well established. *In loco parentis*, as a metaphor of teaching, can trap us all, and not surprisingly, given that students come to us—and we go to them—not far removed, temporally, from our own parental and familial relations (Grumet, 1998). The elaborate psychic processes of idealization, splitting, persecution, projection, and negative and positive transferences constitute and confound the work of teaching and learning. There is no altruistic position, no position outside the trappings of the psychic and the social. Insisting otherwise reenacts a negation that preempts a grappling with the complex ways in which our psychic dramas (and those of others) are implicated in our teaching (and learning). But what seems most significant about the recognition of the psychic dimensions of teaching and learning is expressed by Sharon Todd (1997) when she notes that "affect[ive dimensions are] not simply an individual or idiosyncratic feature of pedagogical life but [are] structurally operational in what gets learned, by whom, and how" (p. 5). For that reason, the shift to "implication rather than application" (Pitt, Robertson, & Todd, 1998, p. 3) heralds a radical self-reflexivity in teaching and learning.

Britzman (1998) notes that transference, understood as these (re)new(ed) editions of old conflicts, is inherently "ambivalent, invoking both unresolved conflicts and profound desires for love" (p. 41). But, as a site on which to reconstitute and to replay psychic (and familial) dramas, teaching is also a site of impossible love, even an impossible site of love, one that offers oscillating and often extreme forms of acceptance and love and resentment and rejection. Ambivalence is the lens through which one views and is viewed. It not only manages one's impulses to hate; it also manages trust, the extent to which one will love and (believe oneself to) be loved. As I go to teaching, always, in the name of an other, yet never quite separate from this other, my conflicts and my ambivalences frame who and how I am as a teacher. My feelings about teaching, grounded as they are in this mother-daughter relationship, resonate on other related sites: teaching is something which carries with it both joy and dread; it is something I cannot abandon yet never stop wanting to leave; it is something I love but sometimes do not want to do. My feelings toward it are, in a phrase, highly ambivalent—a two-step waltz in which the dance partners are separation and connection. The loss of my mother brought these ambivalences (and their psychic sources) into high relief, but their fuller confrontation would require a change of setting, a confronting of another ambivalence, that of return.

The Place of Reparation

A map representation of the island of Newfoundland recalls the shape of a closed hand, a fist, with folded fingers and thumb and the index finger pointing slightly northeastward. In many ways, this image is a fitting one. On the most easterly edge of Canada, lodged defiantly in the North Atlantic Ocean, its dramatic shoreline of sharp cliffs and jagged rock suggest the harsh severance from the main that was its creation as an island. A cultural mythology of struggle, attachment, protection, and survival is at one with this geography. The grip of the place on its people is as the grip of the sea on it: fierce, passionate, unrelenting, haunting. It is a place that, through wind and wave, can toss you around but, through historical affect and attachment, does not easily let you go. As with an encounter with a dramatic persona, it is unforgettable, just as it is a place that does not let you forget. In a novel brimming with vivid, sensual, poetic evocations of this place, the Pulitzer Prize–winning author E. Annie Proulx provides this initial description, through the eyes of her narrator, as she approaches the south coast on the ferry from Nova Scotia, of a community not far from the home of Ann Harvey, the protagonist of *Ann and Seamus*, discussed in Chapter 3: "This place...this rock, six thousand miles of coast... sunkers under wrinkled water, boats threading tickles between ice-scabbed cliffs. Tundra and barrens, a land of stunted spruce... How many had come here...Vikings, the Basques, the French, English, Spanish, Portuguese. Drawn by the cod, from the days when massed fish slowed ships on the drift for the passage to the Spice Isles..." (pp. 32–33).

Later, in a simple, exact(ing) phrase, she captures what many describe as a central characteristic of Newfoundland: "a strong place" (Proulx, 1993, p. 34). It is also a place that elicits strong emotions (of love and hate), a place of great paradox and contradiction, at once somewhat insular and stubbornly separate, yet also inviting, open, and caring—a condition borne, in part, of a history of struggles, politically, with various forms of separation and connection.

My relationship to this place, like that of many others, has always felt spiritually bound, and much less about family and community (although threads of these relationships are part of my relationship to this place) and much more about the drama and power of the geography itself. In many ways, the geography of this land *is* the geography of the psyche. Newfoundland is a place I have always called home. But what does it mean to call a place home? If social relations take root in place, what sets of relations are conjured in calling Newfoundland home? Philip Sheldrake remarks that "place is always a contested rather than a simple reality [and] the human engagement with place is a political issue...because the way it is constructed means that it is occupied by some people's stories and not others" (Sheldrake, 2001, p. 20). In this sense, the description provided in the opening paragraph of this section is both connotational and relational, as much about history, politics, and culture as it is about geography. What might an exploration of these issues of the politics of engagement and attachment reveal about the relationship of teaching and place? As many cultural producers—writers, visual artists, filmmakers, and others—have demonstrated, Newfoundland and Labrador is not without its social anxieties and repressions as well as and along with the communal capacity to offer healing opportunities. *Sense of place*, when constructed on a series of denials and repressions that are shrouded by affirmations of pride and loyalty but symptomatic in expressions of nostalgia, defensiveness, and introversion, can smother efforts to name what has been denied and repressed. Such psychic maneuvers foster a protectionism that can stunt growth and discourage any substantial change.

In his insightful work on the social psychoanalysis of place in relation to the American South, William Pinar (1991) points to the presence of the past in the social subjects of the South, the inhibiting effect of presentism, and the need, culturally and personally, for Southerners to reexperience suppressed pasts in order to enable growth and renewal. Likewise, Marcelle Christian (2001) discusses "the psychological splitting of America" following the attacks on the World Trade Center on September 11, 2001 and the problems of a world constructed "by those who are for us and those who are against us." (p. 12). She, along with others (Butler, 2004), points to the national need for "a conversation about our own less-than-honorable impulses" so that a more nuanced view of the American nation might emerge, one that would encourage a more complex and compassionate view of others (p. 13). Both Pinar and Christian point to the importance of seeing places and cultures psychoanalytically. Teaching infused by a sense of place, then, necessitates a radicalism that preempts such conservative impulses borne of a critical pride and defensive regionalism or nationalism. It asks of its subjects how an emotional history of place is constituted, how its attendant loyalties might betray repressions and denials, as well as love, and how our attachments to and struggles with place reiterate other psychic and social relations

Much has been written about disabling, nostalgic notions of home as enduring (unchanging) presence. Klein herself wrote of relationships with homeland as part of her discussion of *displacement*, the process by which love is externalized to objects other than the mother but that, simultaneously, stand in for her. Klein's arguments point to the psychic depth of these attachments and provide insight into their persistence—and importance:

> The process by which we displace love from the first people we cherish to other people is extended from earliest childhood onwards to things. In this way we develop interests and activities into which we put some of the love that originally belonged to people....By a gradual process, anything that is felt to give out goodness and beauty, and that calls forth pleasure and satisfaction, in the physical or the wider sense, can in the unconscious mind take the place of this ever-bountiful breast, and of the whole mother. Thus we speak of country as the "motherland" because in the unconscious mind our country may come to stand for our mother, and then it can be loved with feelings which borrow their nature from the relation to her. (Klein, p. 333)

Several of Klein's examples of love displacement included in her essay allude to place. For example, Klein argues that the explorer, in the work of charting new territory, reenacts early psychic patterns of an escape from and, simultaneously—through new places—a rediscovery, preservation, and renewal of the relationship to the mother. Similarly, she explains stubborn loyalty to places that are harsh and destructive as enacting comparable patterns of preservation (of place as mother). In such reparative instances, staying (in a place), despite its unforgiving, ungiving nature, can be equated with preserving and maintaining closeness to the mother, while leaving (to explore) can be equated with a quest to find a new, more (for)giving mother. In my own case, struggles with separation from place had the emotional layering, intensity, and resonance of other, earlier struggles. But struggle, by its very nature, is a state of ambivalence. In light of the ambivalence of struggle, what, then, might returning to this home place mean?

As the discussions in the previous chapters have demonstrated, migration is more than a movement of people from one *place* to another; it is also a moving from and through one set of social and cultural relations into another. Herein, place is definitional, constitutive. Following the twentieth century, which brought more migration than at any other time in world history, as a new century unfolds, we continue to grapple with the implications of movement, temporality, in-between-ness, fluidity, transition, and time-space compression. Metaphors of hybridity, creolization, and improvisation capture the mark of movement on and in identity as diasporic peoples (intra-, inter-, and transnationals) negotiate displacement and loss. For example, Dionne Brand (2001) challenges the nomenclature of migration in relation to an African diaspora created by the rupture that is slavery:

> Migration. Can it be called migration? There is a sense of return in migration—a sense of continuities, remembered homes—as with birds or butterflies or deer or fish. Those returns which are lodged indelibly, unconsciously, instinctively in the mind. But migrations suggest intentions or purposes. Some choice and, if not choice, decisions. And if not decisions, options, all be they difficult. But the sense in the Door of

No Return is one of irrecoverable losses of those very things that make returning possible. A place to return to, a way of being, familiar sights or sounds, familiar smells, a welcome perhaps, but a place, welcome or not. (Brand, 2001, p. 24)

Against this backdrop of movement, strong attachments to place, often articulated in terms of region, nation, province, and state, have stubbornly remained—even as some of these same places, in name, at least, have "disappeared."

Unrelenting attachment to place, or to the idea of a place, can still for all remain as a sign of the repressions and longings that form the basis of these attachments that, when left unconfronted, cannot easily be redefined. As Rinaldo Walcott (2000) notes, separation is conflicted—as it is in child and parent relationships—because of the complex nature of projected feelings of guilt, ambivalence, and longing around such places and the identities they shape and mobilize.

At present, part of what marks the evolution of Newfoundland and Labrador within a global economy is massive outward migration, in the past fifteen years, of upward of seventy thousand people of a total population of only one-half million, a trend that showed its first signs of reversal in the population estimates of late 2007. Often accompanying the difficulties of leaving a place one has, for generations, called home and reestablishing oneself elsewhere is the hope and longing for a later return. How do psychic dramas infuse this hope, often felt as an entitlement, a right to return? What other dramas are enacted against a backdrop of the impossibility of return due to war, exile, or environmental devastation? Gayatri Chakravorty Spivak (2000, pp. 344–45) recalls the mythic (and biblical) dimensions of diaspora, the root, in sin and guilt, of such "scattering." What Spivak calls this "ancient diasporic thematic of responsibility and reparation" continues to resonate in (some) contemporary forms of ambivalence toward home land by those who have migrated or left. In this sense, reparative impulses borne of migration echo these ancient mythic and psychic movements. Part of the effect of diaspora is written in the grammar of the psyche as the powerful unsettling of migration finds its register in the psyche through the struggle for separation or renewed connection, the ambivalence borne of a love-hate relationship: the migrant is the child of the mother(land).

In writing about alterations in the mother-child relationship at various points in the child's development, Klein mentions the "store of love" that, in full identification with her own good mother, the mother holds prepared for her (adult) children for whenever they may need it, a safe sanctuary to which they may return (Klein, pp. 319–20). Again, Brand (2001) notes this connection. She asks, "Why do I slip into the easy-enough metaphor of Africa as body, as mother? Is it because the door induces sentimentality? The idea of return presumes the certainty of love and healing, redemption and comfort" (p. 90). Part of the intensity of grief I felt on my mother's death was this loss of sanctuary. In the full throes of grief, I could not believe it gone; life without it was incomprehensible. My return to Newfoundland two years *following* the death of my mother, I believed, would confirm this loss I could not fathom or realize. Instead—and as well—my return affirmed *the place of reparation*, a place in which I could begin the important project of mourning, of staring down my illusions, and, in so doing, beginning the larger project of remaking the attachments that had, for a lifetime, stood in for love. It is, then, an uncanny twist that part of what I now realize my return to this place enables me to do is to love and to feel love(d), to fear and to find solace, to (be) hurt, to forgive and to be forgiving—in short, to make reparation, to shore up a belief in the existence of a world that can be good and just so that I might (re)commit to the imperfect work of educating toward this vision. Being here in this loving place helps me remember what, in grief, I had come to forget as a deliberate refusal to connect so as to avoid further pain of loss: to educate is to reach toward goodness, to encourage meaningfulness—through and in all its difficult detours. In this sense, teaching *in this place* has become a deliberate practice of reparation, the rebuilding of reparative meaning out of the void of grief.

In what is a critique of the insight that Klein provides, Doreen Massey (1994) rightfully points to the potentially regressive nature of attachments to homeland, of "views of place as Woman, as Mother—as what has been left behind and is (supposedly) unchanging." She comments that such a view "is comforting but it is to be rejected. Places change; they go on without you. Just as Mother has a life of her own" (p. 230). Yet, rejection of constitutive ideologies and, and as, attachments is neither easy nor simple, any more than is separating from parents. The psychic complexity of attachments of any sort must be understood before they can be dismantled and one might conceive of possibilities for progressive separation-as-attachment. In the axiom, *you can't go home again*, both the traveler and her home are fluid, changing elements. Yet, it is more readily acknowledged in the longing to go home that it is the traveler who has changed and who can no longer fit (back) into the sets of relations which (once) were "self at home." But it is possible *to go home again* if one's goal is to build so as to live within a new set of relational dynamics, one that refuses the *myth* of oneness, security, and belonging for a committed *ethic* of responsibility, change, and interconnection (Massey, 1994, pp. 110–12). This return—to confront rather than to recover a past and to participate in building a new future home—requires a mourning of old, regressive relationships (be)for(e) the adoption of new progressive ones—here or elsewhere.

But return may be an often-desired but still uncommon or even impossible part of the migrant journey, and, when return is possible, the choice to return is often (and usually) yet further cause for displacement. Actual and nostalgic relations intersect upon return. The actual cannot live up to the highly edited, flawed creation borne of the intersection of loss, longing, and memory that is nostalgia. But these nostalgic images, while not entirely true, are also not entirely false. In mourning, I found myself searching for the very reassurance Klein identifies—the existence of good objects through a community of remembrance of the one lost. I could not find this community where I was, so, in this sense, return held an imperative, prompted by a need to rediscover and (now) more adequately respond to a source of love that has cradled me at various points in my life: a reparative (re)turn. My own return to Newfoundland has prompted a coming to terms with (some of) the conditions of self-change following years of focus, through teaching (elsewhere), of encouraging change in others, my students. In teaching, now, as always, love and place (re)configure social and psychic designs for learning. But, in the name of love, home can no longer compel me to stay or to go. The pulls and pushes of this place are multiple, complex, like the place itself, like me, like anyone. And my ambivalences are now creative wellsprings, sources of critical posturing and agenic lea(r)nings.

As was my mother, and as both presence and absence, Newfoundland and Labrador had always been strongly implicated in my teaching. My early understanding of education as political was in my work as a high school teacher committed to teaching as a vehicle for cultural esteem and celebration and as a means to enact history to build resistance and encourage change. This commitment continued in my work as an academic away from this place. It seems less odd to me now that my identity as a Newfoundlander was disrupted more by my return than my departure. Being away, I clung to that which I had left or abandoned, as I also reinvented it. My return was a watershed of grief, a mourning that shored up both my connection to *and my separation from* family, from place, and from the ways of knowing, the epistemologies, on which old connections to these things were founded. The reparative return to Newfoundland created the space for a sea change, a healthier perspective on the possibilities of connection, separation, and reparation.

Teaching as Reparative

Wounds need to be taught to heal themselves.

Jeanette Winterson, *Art Objects*

What are the implications of understanding the work of teaching as reparative? How might an understanding of Klein's notion of reparation as the development of particular forms of consciousness, as *conscience*, be useful in teaching? Through a consideration of such questions, my own commitment to education across social and cultural difference is problematized herein. In retrospect, my reparative urge, to contribute to social change through education, was fueled, in part, by the urge to contribute to changing the conditions of the lives of others, particularly women, because, when I was a child, it was not possible, through any efforts or achievements, for me to change the conditions of the life of someone I loved—my mother. And while I was able to change some of the conditions of my life, I could not live them freely, without ambivalence and guilt. If loss-as-regret is bequeathed, its inheritor accepts it guilt-wrapped, inviting reparation. But, what seems most important about such a legacy of loss is what is done with it as and while one is discerning what it makes of us—for such bequeathals are not deterministic; within a legacy of loss we are always both inheritors and reinventors. Even if not clear sighted, reparative urges, as Klein points out, are creative and dominated by love, and they are greater than ourselves, connecting us to others in caring ways.

In my own work in teaching across social and cultural difference, understanding reparation now strikes me as an essential ingredient in being vigilant and guarding against and avoiding, in contemporary struggles for change, a new edition of the colonial missionary position. In the pursuit of such vigilance, the suspicious character of ambivalence is a resource. As Megan Boler points out, "suspicion is linguistically active" and "a beginning critique of *regime de savoir*" (1997, p. 236), in our own personal narratives, as well as in those larger narratives not separate from our own, that organize and regulate social, cultural, and educational practices. Noting the sources of our reparative tendencies is an issue that warrants further exploration beyond the confines of any one autobiographical analysis. Jen Gilbert, in her discussion of colorblindness and antiracist education, asks, "How will we know whether our attempts to make reparation with others are working through the dynamics of inequality?" (1998, p. 33). Reparative urges in education have often resulted in the bad effects of good intentions. As a response, Deborah Britzman poses a reminder of the premise on which this discussion of teaching and reparation began: "It behooves educators to engage in the making of reparation that begins in the acknowledgement of their own psychic conflict in learning and how this conflict is transferred to pedagogy" (1998, p. 134). In this regard, the questions of what brings one to teaching and what one brings to teaching are profound questions of educational implication, the implications of an education understood as psychic event.

As well, specifying the context of reparative teaching—what it means to teach across social and cultural difference in a particular place imbued with local and global forces—*in this place*—raises issues about the particularity of this history that has not been confronted and has not been mourned effectively and, as such, might constitute what Pinar calls "a curricular provocation" of such repressed history (Pinar, 1991, p. 177). In my teaching in Newfoundland and Labrador, questions about the ambivalent nature of our relationship to imperial England, on the one hand, and "the Irish diasporic family," on the other, constitute one point of analysis and our place within the Canadian "family of provinces" another. The conflicted histories of divisions and denials around gender, race, social class, and sexual identity, as they relate specifically to this place and its provincialism, are among others. Lisa Appignanesi says, "We all internalize the discourse of the master, the colonizer, the aggressor. Jews, Blacks, immigrants—all carry within them that little nugget of self-hatred, the gift of the dominant culture to its 'lesser' mortals. At times, the nugget is dusted off, polished into brilliance, transformed into pride, brandished on communal occasions. But it rarely altogether dissolves. And it retains a bitter aura of shame" (1999, p. 35). How forms of hatred are learned, as culturally specific and systemic, is a broad-reaching question, for all cultures and all places, the answers to which must be responsible to a new future for any place.

In her discussion of *idealization*, Klein points to the intensification of love-hate in relation to one's parents during adolescence and, concomitantly, the development of highly idealized and highly vilified (even demonized) figures, figures of love and figures of hate. According to Klein, idealization works to reassure one of the existence of good. Vilification, or hatred of certain others, she claims, works thus: "The division between love and hate toward people not too close to oneself also serves the purpose of keeping loved people more secure, both actually and in one's mind. They are not only remote from one physically and thus inaccessible, but the division between the loving and hating attitude fosters the feeling that one can keep love unspoilt. The feeling of security that comes from being able to love is, in the unconscious mind, closely linked up with keeping loved people safe and undamaged" (Klein, 1975, p. 330). In other words, just as love is projected outward, onto others, so too is hate. The intensification—galvanization, even—of social and cultural hatreds (homophobia, racism, sexism), as projections can be partially understood through Klein's insights. A deeper understanding of love and reparation and hatred and persecution can begin to help educators understand and address such social and cultural dynamics beyond the limited educational reach of superficial, rationalist-based attitude adjustment. In her discussion of antiracist pedagogy, Deborah Britzman warns against the educational folly of a "reliance upon cognitive content as a corrective to affective dynamics" (Britzman, 1998, p. 111). Not only does such reliance limit our understanding of the intensity of forms of social hatred and widespread tolerance of inequities, "nor does this reliance on cognitive correction reach the more difficult question of how internal conflicts fashion and attach to discourses of hatred" (p. 112). As Klein and others have established, hatred is an aspect of the psychic dynamics of love, and its historical markings on identity, left unchecked, have far-reaching implications.

Reconciling the struggles and torments of identity are at the heart of a radically reparative pedagogy. Rinaldo Walcott uses Klein's notion of reparation in a discussion of creolization, hybridized and nonessentialized forms of identity created through historical displacement and struggle. A creolized pedagogy is one in which the complexities, ambivalences, and (re)inventiveness of identity—in particular Black and African identities in relation to slavery—might be recognized, analyzed, and interrogated. Walcott argues that creolization "offers a way of seeing defeat as more than lost/loss. It is a practice that may impede the transformation of defeat into (our) shame" (Walcott, 2000, p. 139). Central to this work of confronting productively the presence of repeated violent trauma in relation to one's identity, Walcott argues, is the psychic process of reparation, "a way to come to terms with the rupture that produced a Black diasporic population in the Americas" (p. 148). Confronting the "love/hate relationship of the Americas...is [a way of] making reparation to ourselves, an acceptance of the ways in which the pain and the pleasures of history have intervened to invent us" (p. 149).

Walcott's work is a compelling example of a pedagogy of reparation, a pedagogy that encourages a loving kindness toward the contemporary effects of the slights and damages of colonial history. He emphasizes, in concert with Klein, the importance of self-love and self-acceptance (of a self fragmented, creolized, hybridized) as a precondition of ethical relations with others. Problematic attachments, no less deeply felt by being so, can be negotiated and changed in the space created to offer a loving gesture to oneself. In this (loving)space is also created the opportunity to form new attachments to old sources of love, attachments that bear the mark of responsible engagement for change. But for such work to be successful, the objects of our displacement must be clearly scrutinized for what they tell us about how love has both helped us *and* hindered us. Walcott writes of the diasporic Black and the haunting issue of allegiance to "Mother Africa," noting that separation neither demands guilt nor repression of the historic conditions of diaspora (pp. 147–49). While it is important not to lose site of the specificity of Walcott's arguments as they relate to diasporic Africans, Walcott's point also bears some relation to the experiences of migrant groups other than Africans. The broader points of separation, guilt, and repression—and the subsequent role of reparation—can be more generally and aptly applied to an analysis of the dangers and possibilities of new, emerging, and embedded identities.

New Reparations:
Grievous Lea(r)nings

Grief moves us like love. Grief is love, I suppose. Love as a backward glance.

Helen Humphreys, *The Lost Garden*

In an essay framed around the psychoanalytic notion of matricide, Alice Pitt discusses the problem of representing the mother in teaching and learning and the still persistent and widespread erasure of or ambivalence toward intellectual mothers—women theorists and philosophers. Pitt (2006) notes, "We atone for our matricidal acts [of nonrecognition] by attending to traces of influence that take us by surprise and shatter our illusions of originality. This [atonement] is the work of thinking back *to* our mothers. By doing so, we learn a great deal about learning, and we may also be able to stem the tide of forgetting" (pp. 103–4). While Pitt is not writing here about birth mothers, my effort to write herein about the emotional and intellectual influence of a mother on a daughter may be seen as an ongoing effort of my own to remember as reparative gesture.

The disruption that is the (re)enactment of loss through death created the conditions through which I was forced to examine some of the ways in which three sources of inspiration and love—my mother, my home place, and my teaching—were of a (psychic) piece. The ongoing (re)considerations that form the basis of this recognition are part of what it now means for me to explore, differently, the place of love—and its vicissitudes—in teaching. If love has been the acritical "surround sound," the heartfelt core of a story of teaching, then grief has been its deciding, decided aside, the parenthetical comment, the footnote to the story. Just beneath the surface of my own passions—my love of teaching, my love of place—and masked by a striving for competencies, achievements, and successes, was there a daughter, seeking reparation for the mistakes of identity, still longing to be loved, and heartbroken at the seeming impossibility of it all? Was this the limit of longing? If so, this reparative love had been the thread of connection through my efforts "to do justice." The disorientation and chaos created by loss threatened to break this thread and, thus, imperil all that it sustained, including my teaching. It was little wonder I felt I had lost my way, for in grief we do lose our way. But rethreading a renewed commitment to teaching and place, through remembering the contributions of my mother to who I am, to the work I do, and to the love I gather onto me, has dissipated the threat.

Teaching and learning will reach into and wrestle with these enactments of love, creating the conditions for understandings that might render possibilities for a meaningful emotional life freer from the strictures of our histories and attuned to the need for reparative gestures toward oneself and others. It is likely that, in our teaching efforts, it is possible to only ever approximate the level and quality of caring such education demands. Its complexity—part of a pedagogy of reparation, or understanding teaching as reparative—requires more than can be summoned by a call for (more) caring in education or a reinsertion of what can sometimes be regressive notions of the maternal in teaching. Teachers who understand that the basis of a good relationship to others, for both students and teachers, is a good relationship to oneself, who acknowledge the psychic complexities of teaching and who bring an informed, loving, and reflective presence to learning environments embody key elements of teaching as reparative. A certain carefulness about the emotional life that can be recreated in the relational work of teaching and learning is needed, a carefulness that provides space for a meaningful examination and redefinition of the habits of experience. Proceeding in teaching with such an attitude of carefulness requires a lens that might enable us to examine what our pleasures and pain, our anxieties and antagonisms, our refusals and reprisals tell us about the psychic life that constitutes us as teachers and learners—and the grief and grievances out of which they can be borne.

If loss is our common human denominator, how troubling it is that it is also the emotional place where so many have become least articulate and most afraid. This broken bond must also be repaired, as part of making reparation to our greater humanity. An education that acknowledges, honors, and endures the vicissitudes of love and loss is an education that will sustain life and heal the ravages of difference. But it is not possible to participate in an education attuned to loss without acknowledging the place of loss in one's own life as the basis on which to encounter the losses of others, those who also participate in and are shaped by our educational projects. Paulo Freire (1970) captures this healing power of loving dialogue:

> Dialogue cannot exist, however, in the absence of a profound love for the world and for people. The naming of the world, which is an act of creation and recreation, is not possible if it is not infused with love.... Because love is an act of courage, not of fear, love is commitment to others. No matter where the oppressed are found, the act of love is commitment to their cause—the cause of liberation. And this commitment, because it is loving, is dialogical. As an act of bravery, love cannot be sentimental; as an act of freedom, it must not serve as a pretext for manipulation. It must generate other acts of freedom; otherwise, it is not love. Only by abolishing the situation of oppression is it possible to restore the love which that situation made impossible. If I do not love the world—if I do not love life—if I do not love people—I cannot enter into dialogue. (Freire, 1970, p. 69)

A belief in deep mutuality and an acceptance of the interdependence of life shape the notion of the loving dialogue of which Freire writes. Eve Kosofsky Sedgwick offers one way to realize this mutuality. Toward the end of *A Dialogue on Love*, her courageous and moving account of psychotherapy following treatment for cancer, the distinguished scholar and teacher describes her practice of a Buddhist meditation for connection and care. The meditation involves looking lovingly in order to recognize in each face—whether in a classroom or a crowd—a caring presence, that of your mother in another life (and the child you may have been to that mother). In practicing this meditation, Sedgwick tells of learning "to find [in every face] the curve of a tenderness, however hidden. The place of a smile, or an intelligence" (Sedgwick, 1999, p. 217). Carrying this loving recognition into her classrooms while acknowledging that, she, too, figures in the relational dramas of her students, Sedgwick is repositioned in her teaching to see the intricate and lifelong relationship of teachers and learners. Through such exercises we teachers can practice seeing ourselves and others differently, more compassionately, in connection, and with love. When, as teachers, we learn to grapple with our implication in what Sedgwick (1999) calls the "interminable remediations" (p. 169) that constitute the psychic lives of students and teachers, we may come some way toward radical reparation, toward making better on the various injuries education can compound in the name of learning.

POSTSCRIPTUM: IN ABSENTIA

Outside, an emerging spring; inside, the bustle and pomp of ritual and convocation—it is a day befitting these black-robed graduates whose youthful energy fills the auditorium. She loved this day. And in another, different life, she could have been here, not in the audience, as she once was—a proud parent—but here, where I am now, robed, a part of the procession, the academic order. I remind myself that it is some small recompense, that her name will live on, even if she could not, that her name will now be attached to one of the finest of these young women, leaders in teaching and service, like she was. The heat builds beneath my red, red robes—stifling academia—her gift to honor my achievement. My hands are clammy now, following the order of graduates through the program, waiting for this moment when the inaugural winner of this teaching prize begins to cross the stage. She is so young, yet already she embodies a quiet dignity and composure, like she had, too. The hooding comes now, as the award is announced and my mother's name fills the theater, as it will hereafter, on this convocation day: an interminable thank you, presented *in absentia*.

Conclusion

"Learning to Live with Ghosts"
Loss, Place, and Education

A Conclusion

Together, the chapters of this book reiterate an argument for a more concerted focus on loss in education. Cultural crises—of which loss is an indelible feature—demand a rearticulation of educational vision at local and global levels, as is well demonstrated in the work of many contemporary scholars working on many different fronts. The work of Edmund O'Sullivan (2002) in transformative education in the context of the threat of planetary collapse, that of Megan Boler (1997) in critical emotional literacy in the context of unrelenting imperialist war, and that of Rachel Kessler (2004) and her PassageWorks Institute (2000) devoted to the inner life of young people are but three examples that speak to some of these (not unrelated) contemporary crises.

The visions of these educators are in stark contrast with prevailing notions of education, those that shape and maintain current technicist, rationalist, and decontextualized forms of education. Such forms of education are effective in maintaining the ill-based logic of exploitive economics, rampant consumerism, and unsustainable communities. However, they are not befitting a context of planetary depletion, increased global conflict, and unprecedented cultural destruction. Many educators realize the incongruities between their immediate, prescribed tasks at hand and the urgent needs of their students and communities. Such incongruities create great difficulties for educators as, increasingly, the limited possibilities of dominant modes of education are confronted. Such lost ideals of teaching and learning are not unrelated to the crises an educational discourse of loss might address (Phelan, 2003).

In each of the preceding essays in this book, I have addressed aspects of a cultural crisis particular—but not limited—to Newfoundland and Labrador and some of the personal, social, ecological, and psychic dimensions of this crisis. In so doing, I have gestured to a notion of education that challenges accepted traditions within mainstream educational practices and, as well, but to a lesser extent, within critical practices of education. In this concluding chapter, while drawing from these preceding discussions and returning to issues outlined in the introduction, I attempt to highlight briefly what I consider to be key characteristics of an education that addresses cultural crisis, loss and change.

Avowing Life:
The Acknowledgment of Loss

Loss is the persistent condition of life, the very basis of subjectivity (Butler, 1997). Yet, little time is spent in education explicitly wrestling with this fundamental human bond and both its enabling and disabling dimensions. Contemporary times demand even more so that this disjuncture between what unites us and that about which is spoken in education be addressed. Rachael Kessler (2004) argues that we are "a culture afraid to sit with the feelings of loss" (p. 147). Furthermore, as Joan Didion (2005) notes, great social admiration accrues to those who hide grief well in a world in which to express grief is to be considered self-indulgent. Part of what such fear and repression instill is a refusal to acknowledge loss and an "othering" of those whose expressions of grief differ from our own. Without a meaningful acknowledgment of loss—as an individual and, importantly, a cultural experience—a deep and integrated education is impossible. As I indicated through the discussion in Chapter 1, the regulation and denial of loss permeates every level of contemporary culture, sanctioning some forms of grief and loss while denying and dismissing others. Such regulation is managed through institutionally binding silences that numb and dishevel, fragmenting and breaking any potential agency that might accrue from loss and its inception to change.

The acknowledgment of the complexities of loss—who is lost, what is lost—begins the real and arduous work of mourning. Judith Butler (2008) warns of the cultural melancholy that arises from the failure to acknowledge, the disavowal, of loss. For Butler, the critical articulation of grief is an essential component of a sensate democracy (Butler, 2008, p. 3). From this perspective, disavowal not only mutes our capacity for grief and our ability to mourn, it also stymies our ability to participate fully and empathically in civic matters. Culturally pervasive melancholy, nostalgia, and solastalgia, discussed extensively in Chapter 4, attests to the extent to which this incapacity to feel and to acknowledge emotion, in this case the affective structure of grief and mourning with all its attendant emotions, marks contemporary life, including the identity politics of displaced communities and multicultural nations (Gilroy, 2005). Such structures of feeling gesture to a profound wellspring of unresolved loss. If loss is the impetus to remember, then to disavow loss, to refuse to acknowledge loss, is also to mute a capacity to remember and to act ethically in relation to memory. Deborah Britzman calls such an ethics of loss and memory a "learning to live with ghosts." She explains, "Learning to live with ghosts means learning to understand what has been lost in the self and what has been lost in the social. This is a question of becoming an ethical subject in relation to other ethical subjects, a question Freud called 'the work of mourning'" (Britzman, 2000, p. 51). This ethics confirms "the thrall in which our relations with others hold us" (Butler, 2003, p. 23), the mutually constitutive basis of our ties that bind.

Articulating Grief: The Necessity of Mourning

In a riveting musical video entitled *Memory Waltz*, Rawlins Cross (1992), a now disbanded Canadian Celtic rock group, most of whose members are from Newfoundland and Labrador, powerfully demonstrates the reach of cultural mourning and the importance of expressive culture in this work of confronting loss. Against the background of a haunting, elegiac melody, black-and-white images of a once-thriving and now decimated cod fishery and the communities formed around it are intercut with pictures of solemn, funereal-looking band members, attired in black and white, and playing traditional Celtic instruments. The music and images—newspaper headlines, still photos, video footage—narrate the cultural catastrophe that has unfolded over the past several decades. The music video ends with the image of a small dory being eclipsed by a mandolin—a symbolic rendering of the place of expressive culture in the reconciliation of cultural loss or, alternatively, the cultural and melancholic incorporation of the lost object. Like Lawrence O'Toole's memoir, *Heart's Longing*, discussed in Chapter 2, in its heartbreaking articulation of loss and mourning, the music video is a beautiful evocation of—and catalyst for—dialogue about the acknowledgment or disavowal of loss and the necessity of mourning, both individually and culturally, in all its permutations.

Such elegiac creations can help move a culture forward into a clearer, more compelling and critical articulation of cultural loss and mourning. They offer a place from which to practice what I have called a reflective grief and to begin to repair through public dialogue that reaches across difference to a place from which to reconstruct communities and futures. If the disavowal of grief is a mark of contemporary society—in a world where there is so much cause to mourn our fractured and damaged humanity—then the ability to articulate grief atrophies through repeated denial and disavowal. Many contemporary discursive expressions of grief are popularized and formulaic, attesting to the manner in which we culturally learn sanctioned, if not necessarily effective, ways of expressing loss. While these popularized discourses of grief attempt to honor loss, they can often appear vacuous and clichéd, caught up as they often are in the "finding closure" discourses so common in contemporary discussions of loss and grieving. Learning to (re)articulate grief critically becomes part of the project of an education that attends to loss, its dimensions and its implications.

TIES THAT BIND: INVESTIGATING ATTACHMENT

If response to loss attests to attachment, an education that acknowledges loss must investigate the nature of our attachments to people, ideas, objects, and places as part of a project of cultural healing and transformation. As Rebecca Martusewicz notes, "We segment our lives through the creation of specific geographical, social, and personal attachments which never completely leave us. Rather, the traces of these attachments (and detachments) compose us as we engage them" (1997, p. 16). Our attachments are constitutive; they compose us. Investigating attachment is a project of freedom, not an impossible or undesirable freedom *from* attachment but, rather, a freedom from blind, unexamined attachment, at personal, cultural and national levels, and a freedom to attach in meaningful, thoughtful, and reflexive ways.

Blind attachment inhibits our capacity to be human, for it refutes a consideration of what is refused through attachment. It is impossible to be attached to one or in one way without a refusal of another or another way. In this way, detachment is also a form of attachment, an expression of the refusal of attachment that bespeaks powerfully to what is, more so, an attachment—through refusal—to that which is refused. This ethical dimension of an examination of attachment—whether it is called detachment, separation, critical distance, or unbelonging—is necessary, as Martusewicz also argues, so that we may continue to ask fundamental ethical questions about our shared humanity: "How should I be in the world? How do I care for the other?" (Martusewicz, 1997, p. 16). These questions are questions about the nature of our shared vulnerability, our interconnections, the basis on which they are formed, and the forms of human life they allow and disavow. In relation to forms of nationalism—prevalent large-scale, state organized and sanctioned forms of attachment, along with its derivative regionalisms—Judith Butler (2008) speculates that a "critique of nationalism might produce new frames for the human, new ways to read, to hear, to see, and so new possibilities for a sensate democracy" (p. 194) and, in so doing, create new ways to forge the more convivial relationships that would comprise it.

SUSTAINING HOME: AN ECOLOGY OF PLACE

[T]here is a great sadness about what has happened in Newfoundland and Labrador with the closure of the cod fishery, the demise of the inshore fishery, and the massive migration that has followed. You live in place you do not want to leave unless you really must. People here...have a great love of this land. But we know that some of what has happened to the oceans—overfishing, the depletion of fish stocks, and destruction of habitats—is not an act of God. These changes are the consequences of a type of international economy that actually fished without care about replenishment and only with care about profit. (O'Sullivan, 2008, p.xi-ii)

O'Sullivan, "Notes Towards a Transformative Education"

It is a paradox of industrialized society that the very earth that sustains a way of life is not protected so that it may continue to sustain that life indefinitely, for all life forms. It is also paradoxical that strong attachment to place is also too often accompanied by disregard for the needs of a place, as an ecological entity. In such a context, blame-leveling rhetoric and short-sighted, solution-driven thinking about the nature of a particular crisis often prevail. With the implication of all, albeit differently, in the challenges to long term sustainability, yet with few expressions of responsible leadership, and with the impossibility of any immediate quick-fix solutions, the interrelated social, cultural and ecological crises of our contemporary world continue to spiral out of control, often leaving a sense of powerlessness and despair as their most insidious effects. Paradoxically, despair can fuel further damage to the planet when, in avoidance, a turn to the marketplace attempts to fill the gap of despair with a fleeting fix of consumer happiness. John F. Schumacher (2006), writing on happiness, argues that "All worthwhile happiness is life-supporting. But so much of what makes us happy in the age of consumerism is dependent upon the destruction and over-exploitation of nature. A sustainable happiness implies that we take responsibility for the wider contexts in which we live and for the well-being of future generations" (p. 3).

The kind of transformative education outlined in Chapter 4 incorporates social, cultural, and planetary concerns while challenging the virulent consumer base of contemporary logic. Following O'Sullivan (2002), its starting place is a confrontation with contemporary forms of cultural catastrophe as monumental planetary loss. Yet, such a confrontation, while having a global reach, has always an immediate local base of expression and effect. From a context that echoes that of many communities of Newfoundland and Labrador, and similar communities worldwide, Michael Corbett writes,

> I work in a Nova Scotia fishing village where a way of life is vanishing in the face of near-sighted corporatist state policy and technologically-induced over fishing. The struggle, at this point, is to understand what makes educational sense in the face of chronic change, the kind of change that can smash a way of life in a frighteningly short space of time. As difficult as it is to imagine community at this historical moment, in the context of a migrant society in which people are simply expected to get up and go where there is work, my colleagues and I are left wondering about what might count as education for the survivors, those left behind in places history seems to be trying hard to forget. (Corbett, 1999, p. 170)

Any education to address these issues that Corbett raises and to which the preceding chapters have been devoted must begin in the places that we call home and must address the hopes, fears and futures of the peoples of such places as real and realizable projects of a sustainable education of place, one that, as David Gruenewald (2003) argues, interrogates "the links between environment, culture, and education" designed to "embrace place" as "a critical construct in educational theory, research and practice" (p. 10).

Such links surpass materialist inclinations to reside in deeply spiritual, as well as political, modes of attachment.

THE BREATH OF LIFE: SPIRITUALITY AND HEALING

Aboriginal peoples continue to teach that the problems created through cultural loss—in their cases, often, cultural genocide—can only be addressed through profound spiritual healing. The Mushuau Innu people of Labrador are a profound example of such courage of spirit. Natuashish ("break in the river") is a resettlement community built for the Innu of Davis Inlet, a traditionally nomadic people twice previously relocated against their will. In the 1970s, without consent or compensation, a large portion of Innu lands was flooded in the development of the Upper Churchill Falls hydroelectric project. In the negotiation of the Terms of Union of 1949 between Newfoundland and Canada, the aboriginal peoples of Newfoundland and Labrador were not afforded status through the federal Indian Act. Without legal status, access to services, and self-governance, and, until recently, caught between the indifference of two levels of government, the Mushuau Innu have suffered immeasurably. The community is marked by poor living conditions, addictions, high suicide rates, excessive unemployment, and low attainment levels in education and literacy. In 1999, Survival International, a London-based worldwide organization dedicated to the survival of tribal peoples, documented the plight of the Mushuau Innu and, in a scathing criticism of government indifference, referred to the Labrador Innu as "Canada's Tibet" (Samson, Wilson & Mazower, 1999). Natuashish is supposed to be a new start for the Mushuau Innu, but their troubles have followed them there. Some Innu say that "tragedies occur so often that many of our people have forgotten how to mourn."

Speaking of the plight of her people, Tshaukuesh, Elizabeth Penashue, an Innu elder from Sheshatshiu, decries the political indifference that has created the conditions of their struggles to adapt to and to reconcile the cultural life-altering changes brought about by involuntary cultural displacement (Elwood, 1996)—the solastagic effects discussed in Chapter 4 of this book. Speaking about African epistemologies, George Dei points out, "Local communities must be understood not simply within so-called Western rational thought but instead within an indigenous, culturally contextualized genesis" (Dei, 2002, p. 9). An ethics of loss would deem it imperative to listen responsibly to such ways of knowing, maintained despite ongoing and renewed forms of imperialism and colonization, in order to learn through a bearing witness so as to learn to imagine and to live differently. The wisdom of Tshaukuesh and others like her bespeaks not only devastation but resilience, determination, and hope. The Innu know that to heal is a deeply spiritual project that begins with an acknowledgment and an honoring of what is lost. The importance of listening to the wisdom of aboriginal peoples, and of honoring aboriginal knowledge, is paramount in any healing relationship to one another and the earth (O'Sullivan, 2008, 2002).

THE WILL TO RECORD: REPRESENTING LOSS

I think we ought to read only the kind of books that wound and stab us.... We need the books that affect us like a disaster, that grieve us deeply, like the death of someone we loved more than ourselves, like being banished into forests far from everyone, like a suicide. A book must be the axe for the frozen sea inside us.

Franz Kafka, Letter to Oskar Pollak

Like a book that is "the axe for the frozen sea inside us" (Kafka, 1904/1977), expressive culture, now more than ever, must penetrate contemporary pachyderms to enhance the creative power to heal. As such, cultural works form the basis of an education concerned with healing through the production of new narratives of and for new, multiple and contingent forms of belonging.

Many artists and critics have argued that much of the music, art, and literature of Newfoundland and Labrador is marked by a preoccupation with the past, what many have called a nostalgic view of a difficult history. Yet, these expressions serve not just regressive purposes, for they also signal a dissatisfaction with the present (Hutcheon, 1998) and, in their roots in melancholia, an unpaid debt to what is lost, what is past, a debt that can be accounted for through positive, reparative change. But such art can also maintain an affective stasis. In their preponderance, nostalgic expressions also accentuate a dearth of other kinds of cultural expressions that represent and mobilize effective mourning—art that helps us grieve. While this latter kind of art may help us understand nostalgia and its impulses, it does not reiterate or sustain a nostalgic position, nor does it condescend about the toll that is loss and the hard work that is mourning. Effective cultural representations encourage a complex encounter with loss and an acceptance of the difficult contradictions and paradoxes of life, as it also encourages a forging ahead with wisdom and hope.

In *Radical Hope*, his discussion of the Crow nation, Jonathan Lear (2006) writes of the need, in the context of a lost nation of Crow and the end of a way of life, for a new kind of cultural poet. He describes this new poet as "one who could take up the Crow past and—rather than use it for nostalgia or ersatz mimesis—project it into vibrant new ways for the Crow to live and to be. Here by 'poet' I mean the broadest sense of a creative maker of meaningful space. The possibility for such a poet is precisely the possibility for the creation of a new field of possibilities. No one is in a position to rule out that possibility" (Lear, 2006, p. 51). Lear's point transcends the Crow nation and can be applied to any culture facing profound change and to the challenge of its poets, its public meaning-makers, within such a context of devastation and loss to encourage new, more open, and more responsive forms of identity and modes of belonging.

Each of the narratives referenced in this collection attend to loss, albeit differently and, I would argue, with different degrees of effectiveness. There is an emerging body of new expressive culture—painting, books, and music—that engages the complexities of loss in a conscious and abiding manner. In an interview with Leo Furey, Michael Crummey, an internationally published, award-winning Canadian author from Newfoundland and Labrador, comments on the impetus for his work:

For me, I think it goes back to this whole issue of loss. Loss really is my subject, I think. On all levels. There is cultural loss in terms of the extinction of the Beothuck or the end of the fishery, the inshore fishery. Looking at the way that affects a community and me personally.... There's the type of loss that I grew up with in Buchans, for example, where the mines were shutting down and an entire community stopped being what it was and became something different. And then there's personal loss in terms of death, the central experience of human life.... Writing through those is my desperate attempt to cling to most things somehow or find my place in them. Or even just to state the absence for what it is. (2002)

Crummey's words recall those of Sanchez-Pardo: "If the ultimate commandment of History is not to forget, what can be more proper than the will to record?" (Sanchez-Pardo, 2003, p. 393). Since the moratorium on and closure of the cod fishery in Newfoundland and Labrador, this will to record has been echoed by many artists. Susan Dyer Knight, the founder and director of the internationally renowned Newfoundland Youth Symphony Choir, founded in 1992 in the wake of the moratorium and renamed Shallaway in 2006, expresses the project of her choir to "take Newfoundland and Labrador both out and in—out into the world and into ourselves" (Ostroff, 2006) in order to buttress loss through knowing and remembering.

Such a writing and representing must also be an aspect of everyday and less public practices of expressive meaning-making. As Ann Harvey, the central character of *Ann and Seamus*, the verse novel discussed in Chapter 3 reminds, the ability to write *of* home is an essential aspect of mourning loss of place. Lacking this ability herself and fearful of entering a world in which her cultural place was unrepresented, she chooses to stay in Newfoundland. Despite what may be a shared attachment to this place, the contemporary migrant lives in a different world from that of Ann, one in which the ability to represent is culturally demanded through schooling requirements in literacy. It is also a different world in that cultural objects of place—books, art, music, and so on—are readily and ever present. New technologies have also created possibilities for enhanced exchange through such things as Web site blogs and wikis, sites on which nostalgic and melancholic impulses can be shared and debated.

When I was a young migrant who had initially left Newfoundland and Labrador to study, writing and researching my attachment to place were key sites of my own efforts to work through the contradictory longings of home place and, as such, a crucial component in developing my own resilience. What writing also afforded me was space in which to develop a critical posture, a place for rewriting and revising attachments, for reconstituting a life, and, in so doing, it offered an opportunity for a deeper connection to place through a learning to love place differently. Stephen Hatfield, the composer of *Ann and Seamus: A Chamber Opera*, comments, "We have art so that we do not perish from the truth" (Ostroff, 2006). Art—representation—can help us critically confront our cultural truths, testing the mettle of their mythic inclinations and social and cultural assumptions while also providing the solace of connection and dreaming anew. Dionne Brand (2001) writes, "Writing is an act of desire, as is reading. Why does someone enclose a set of apprehensions within a book? Why does someone else open that book if not because of the act of wanting to be wanted, to be understood, to be seen, to be loved?" (p. 192).

Reparation: Hurt-filled Returns

In Chapter Five, I wrote of my own reparative urges through a teaching for social change. If, as Brand suggests, desire is born of dissatisfaction and lack—a longing for the "was," the "not yet," or the "is" that will not be—and change, in this context, is an urge for an imagined, less injurious or lacking state, then the desire for change is an acknowledgment of loss or its anticipation. Teaching for change, then, is a teaching propelled by the desire to reduce injury, to repair harm, and to reconcile loss. In this sense, such teaching can be seen as fundamentally reparative.

Paula Salvio (2006) writes about the possibilities of teaching as reparative through her examination of the pedagogy of Anne Sexton, the renowned American poet, whom she describes as "a teacher engaging in the study of problematic attachments" (p. 83).

Focusing on the character of ambivalence and its relation to melancholy and reparation, Salvio notes, "The work of reparation unfolds on the other side of hate and loss. It calls upon us to live within the tension of opposites: love and hate, anxiety and composure, desire and responsibility, the will to hide and be known, to create and destroy" (Salvio, 2006, p. 68). Living and teaching (amidst) these tensions, as Eng and Han (2003, p. 366) argue, "requires a public space where these conflicts can be acknowledged and negotiated"—an educational space attuned to the vicissitudes of teaching and learning and the dynamics of difference. It is a space in which "to achieve a history that can doubt itself (Walcott, 2000, p.150) as a means by which to confront pain and loss so as to grow and change.

For the migrant, whose identity is marked by the in-between-ness of which Bromley (2000) writes, reparation calls forth other hopes and possibilities, as it also poses educational challenges. These reparative returns, as I discussed in both Chapter 1 and Chapter 5, are opportunities to address difference and the social and cultural conditions under which they are forged and fought. Boldt and Salvio (2006) comment on this in-be-tween-ness and its educational challenges. They suggest that for those who have been displaced, whose identities have been disrupted, and for whom belonging is always problematic and illusive, there is something more required by way of educational response: "There is a vision beyond homecoming, of return, but a return that includes a consciousness of the past, and a commitment on the part of those of us who have inherited the imperialist impulse, to make reparation. To address these inheritances is among the most profound challenges we can take up as educators" (Boldt & Salvio, 2006, p. 110).

Reparation comes out of our best instincts to love and be loved, to reduce injury and to seek solace and peace. It is both irreducibly creative and caring and is the agenic core of our best efforts to educate across social and cultural difference, to educate for social change.

Dynamic Interspecificities: Of, For, and About Difference

The particular, local, and micro interface with and have a mutually constitutive relationship with the general, global, and macro. These dynamic *interspecificities* form "the web of economic, political and ideological domination and exchange" (Boler, 1997, p. 227) that is globalization. As Boler notes, while often cast in metaphors of homogenization and the erasure of difference, globalization connects disparate places, converges space, and is always felt in deeply specific ways in specific contexts shaped intricately by history and its circumstances. Education must attend to the dynamics of such specificity and difference and the points of vulnerability and hope that it galvanizes. Madeline Grumet (2006) argues that education privileges the separation on which difference is founded. "That space between the me and the not-me is not really a space that we have tried to bridge in education, for all our talk of community. We have focused on separation" (p. 215).

The changing conditions of the twentieth century—propelled by unprecedented worldwide migration and postcolonial thinking—have helped create what Kobena Mercer (2000) describes as "a condition of multicultural normalization" (p. 234), but one not registered in the same ways in all contexts. The complexities these variances pose for education across social and cultural difference need to be underscored and not compromised by universalized or globalized notions of the political work needed. While it is important not to eclipse common struggles or to mute strong resonances across different spaces and positions, it is also important to highlight the need to formulate specific questions for investigation and strategies for action that make sense in the historical contexts in which they are framed and formed and that arise from the deep particularity of such contexts.

In the context of global migration, much of it forced by brutal regimes, genocide, horrific impoverishment, and environment devastation, migration cannot be conceptualized, politically, in a monolithic way. Interprovincial migration within a nation state differs greatly from emigration and immigration, and within each and every one of these movements resides great multiplicity and irreducible specificity. Displacement, while injurious, cannot also create infringement and exclusions, reiterating and re-creating imperialist power relations on new sites. Yet, the intricate and ever-present traces of colonialism and imperialism connect these disparate movements when history is reread and realized as partial and political story and where memory is enhanced and revitalized in more intricate, varied, and inclusive ways. For example, in my own early education, I learned of the lost tribe of Beothuck but not of their genocide, that "cloud of bone" (Morgan, 2007) that continues to hang over our culture. Like Ann Harvey of *Ann and Seamus*, while I lived and learned the harsh cruelties of a culture divided along lines of class, gender, geography, and religion, I did not learn how to challenge its constitution. The Innu of Labrador were largely absent from my formal educational map, as were the Mi'kmaq of Conne River. And nowhere was I taught of other historical injustices such as the so-called head tax placed on Chinese immigrants, imposed in 1906 and not lifted until confederation, or of the denial of many immigration applications from European Jews fleeing the Holocaust, or of the existence of an internment camp at Pleasantville Base in St. John's during World War II or of the African slaves who fished the Grand Banks. These gaps and silences persist as part of a curricular repression inseparable from a politics of place that can maintain a separation from broader historical and global developments.

In the face of such continued systemic disavowals, our relationship, as individuals and as a culture, to these injuries of history, remains largely unchallenged, encouraging a deep melancholy. Such melancholia is a symptom not only of disavowal but also of inability, the latter pointing toward structural and systemic obstacles to attachments to new objects and new dreams (Eng & Han, 2003. p. 352), the implication of social and cultural differences in a thwarted reconciliation of loss. Also thwarted in such a lack of reconciliation is any project that might lead toward what Paul Gilroy describes as a "multicultural ethics and politics premised upon an agonistic planetary humanism capable of comprehending the universality of our elemental vulnerability to the wrongs we visit upon each other" (Gilroy, 2005, p. 4).

Against the Grain: Cultural Critique and Uncommon Sense

Appeals to common sense, as to tradition, have a particular cultural place in the regulation of thought and action. Unraveling the political intricacies of contemporary social life means, minimally, that "one has to be willing to complicate what appears evident or straightforward" (Salvio, 2006, p. 69). Such a project of complication requires readings of the affective, social, and textual worlds that are both intricate and profound. Echoing Gilroy (2005), Allan Luke talks about reconceiving literacy as "a tool for conviviality" (2003, p. 1), a means by which to engage—to *read*—a complex world in complex times so as to address "'the imperative of learning to live together ethically and justly" (p. 1). This point resonates with Madeline Grumet's argument that reading is a means by which to refuse separation and to connect and to attach. Reading, she says, "invites us to recuperate our losses [through separation]. As we enter into the fictive world and emerge from it, we experience the opportunity to reconsider the boundaries and exclusions that sustain our social identities" (Grumet, 2006, p. 221) In many ways, this imperative requires that literacy turn its back on its long history of implication in the reiteration and sedimentation of social inequities (Kelly, 1997). As Luke and Carrington note, "Literacy and by association literacy education are both historically constructed and historically constructive, normative enterprises. In current conditions, they are about the shaping of patterns and practices of participation in text-based societies and semiotic economies" (2004, p. 53). As I argued in Chapter 3, the relationship of literacy to social formations and social transformation must be troubled so that its freeing potential may be more fully realized. Part of this project entails "a refashioning of literacy as a normative preparation for a critical engagement with glocalised economies" (p. 64), whereby literacy is "a means for building cosmopolitan world views and identity: of enhancing, in [Pierre] Bourdieu's terms, historical memories and contemporary understandings of how these economies of flows actually structurally position (and perhaps exclude) one, how differing dispositions will have different effects in the various fields of flows, and how to actively engage with those fields in agentive and transformative ways" (Luke & Carrington, 2004, p. 64).

Understanding one's place in such global fields of flow is to trouble configurations of place and its mythic promises. Doreen Massey (2000) writes of traveling regularly with Stuart Hall from London to and from Milton Keynes, where they both commuted for work. She uses the example of their journey to illustrate an argument for "thinking time and space together and thinking both of them as the product of interrelations" (p. 229)—"the simultaneity of space" and "situated simultaneities" (p. 229). Massey attempts to get at "the multiplicity of histories that is the spatial" (p. 231), its "contemporaneity of trajectories" (p. 228). Space does not stand still; it changes with and in time. Likewise, place is not necessarily (or any longer) coincident with community. Geographies can be sites onto which we may manifest our longings; they cannot contain or fully define them. "Home," regardless of where or how one defines it, is, like identity, a highly invested, politically complex mythology. And, like identity, it is a mythology by which we live, and in which we deeply invest, which is why it is so important to understand its emotional dynamics and its cultural politics. To understand the psychic and political conditions of claims to home is to come to know oneself as a nomad, as one who experiences a necessary state of in-between-ness in which one occupies what Roger Bromley calls "a multi-locational imagination" within "a postnational model of belonging" (in Pandurang, 2001, p. 3)—a real yet always provisional place in a project of *unbelonging*. Within such an understanding, insights into the intersections of loss, displacement, and mourning reshape and consolidate a position of ambivalence about belonging any*where*. Not home*less*, but home*free*; the desire "to be home" replaced by the desire to "feel at home." There is no necessary place-factored equation in such a position.

Complementing some of the ideas presented by Luke and Carrington (2004), Paula Salvio (1998) draws on a metaphor of willful world travel, developed by the philosopher Maria Lugones, to argue for a literacy education as engaged thinking, empathic inquiry, and critical practice. For Salvio, teaching is about moving through the worlds and cultures of others, a necessary precondition for which is knowing our own world and culture and the social values on which it has been constructed. Following Lugones, Salvio contrasts compulsory world travel, done for the purposes of survival, with willful world travel, an educative and empathic project of reading, viewing, or moving into other places and spaces in order to begin to see others as full subjects rather than as eschewed objects of what she calls "arrogant perception" (p. 67). Such willful travel can be facilitated through stories of migration and displacement—compulsory world travel—and through active investigations of the "ties that bind" us to notions of place and culture and the identities constitute in relation to them.

There are many public cultural sites on which such stories must emerge, for purposes of scrutiny, interrogation, and reconstitution: testing against the mettle of various discourses and competing claims. Within education, they can emerge as a form of curriculum and, if you will, as method. Migration stories can offer an "affective experience of marginality" (Bromley, in Pandurang, 2001) that can provide an opportunity to engage issues of social and cultural identity, autobiographically, where educational studies become sites of exchange and trajectories for an exploration of alterity, the basis of difference. The beginning place of such study is a tracing of the interrelationship of autobiography, curriculum, and (trans)nation.

It is especially important that such stories emerge within teacher education, in particular. Civic requirements in "these times" include the capacity not just to "border cross" but to understand the political terms and conditions on which any border exchanges or crossings might take place. Regardless of the degrees of diversity in which one's "home" population is constituted, we are all travelers who will encounter other travelers, in home, work, and community spaces, and elsewhere. Such encounters, such "situated simultaneities" (Massey, 2000), both constitute home spaces as they also, simultaneously, challenge this very constitution. Education has a central place in the encouragement of new forms of belonging. This is only possible—feasible, even—when older forms of belonging are explored, understood, and contested. This process is enabled by a trifocal lens that might capture the multiplicity of movements felt in where people have been, where they are "for now," and where they might be going. Part of the work of teacher education is to prepare travelers: those who busily navigate and negotiate the everyday locale, whose stories are always already mobile or mobilized; those whose dreams, desires, and needs will take them to other places or locales as part of a global movement of teaching work; and those whose collective (cultural, familial) lives are indelibly marked by movement or migration.

Inventing Selves: Reel Identities

These times are rife with reminders of movement, of temporality, contingency, fluidity, in-between-ness, and transition. Metaphors of hybridity, creolization, and improvization underscore the mark of movement on or in identity. For some time now, I have been (theoretically) preoccupied with displacement, loss, and longing. As the preceding chapters have demonstrated, these stubborn problematics illuminate and obfuscate, simultaneously, the personal and cultural dynamics that are their embodiment. Emotional, psychic economies proffer the greatest register in struggles to educate across social and cultural difference, and identity claims are at the heart of these economies

I have struggled with the contradictions of dislocation and displacement: the feeling of deep-rootedness of place that exists alongside the knowledge that roots move, grow, can be transplanted, and, when so, often flourish. Roots, I also remind myself, do not just grab into their foundation but move through it, touching, entangling, strangling, separating, yes, but, also, existing symbiotically, enriching. Roots are both beginning places and transitional spaces. They are necessary, life giving, real, and changing, yet they are also confining, disabling, and mythic. Writing about (specifically Chinese) diasporic identities, Ien Ang (1998) notes, in relation to changing and hybrid identities borne of diasporas, that "inventions are not signs of inauthenticity, but only add to the repertoire of modes of being that can be included in the open-ended and polysemic rubric [of identity]" (p. 6). Yet, as inventiveness is celebrated, it is important to note that the capacity for and the conditions and the freedom to assert invention, are not shared equally by all people and do not exist equally in all places and spaces. Part of the work of education is to address such limits and to expand the spaces and conditions of identity.

Newfoundlanders and Labradorians passionately and proudly proclaim their identities. What are considered unfair representations of the place and its peoples often elicit feisty responses. But passion and pride can also be regressive and excluding, as well as strengthening and enabling. Discussions of the desirability of attracting more immigrants to Newfoundland and Labrador elicit various responses. But callers to open-line shows and letters to editors of local newspapers are quick to point out that it is the province's well-educated migrant youth who should be encouraged and offered incentive to return home. While a social color codification may be implicit, if not necessarily conscious, in the framing of this hierarchy, it is also implied that "home" is a place to which some "we" "holds title" (even after we leave) and to which "others" (sometimes even following prolonged habitation) are not (fully) entitled. These dynamics are not peculiar to Newfoundland and Labrador; they resonate in cultural conflicts worldwide (Ang, 2000; Mac Einri, 2001).

Questioning the stakes of and in identity in the face of massive outward migration and increasing immigration, and identity's demands from the shifting and contingent perspectives of "here" and "there" is essential in the forging of new, forward looking identities. Here, the transnationalization of curriculum must be seen within the structural dynamics of the local, the locale, and the glocal, and it still requires attention to what Pinar (1991) calls "a social psychoanalysis of place" (p. 165). These times require identity change and stories that are a reminder that change is afoot. But such change necessitates as much attention to the disorders of identity claims—its afflictions—as its idealizations.

Sensate Democracy: A Critical Civic Emotional Literacy

In her discussion of what constitutes a sensate democracy, Judith Butler states,

> One precondition of democracy is participation. Democracy must be participatory and, at a very basic level, entails a capacity to know the world, to judge the world, to deliberate upon it, and to make decisions that are based upon an apprehension of the world... There is no nonstructured apprehension. But, given that we accept that, the question is, how do we come to apprehend the social and political world?...[W]e have to be able to see images and hear voices, even to smell and to touch a world we are asked to fathom. (2008, p. 3)

The animated civic life of which Butler speaks is not only one that refuses the numbing effects of received knowledge or the Cartesian logic of much contemporary education. It is a form of civility premised on full sensory engagement. Butler sees such sensate apprehension as a precondition of judgment (p. 8) made possible when engagement is combined with a critique of the ways in which it is structured and solicited so as to regulate our perceptual frames—our apprehension of the world. Butler's idea of sensate democracy is enhanced by Megan Boler's notion of critical emotional literacy. Boler (1997) defines emotional epistemologies as ways of knowing that incorporate "a public recognition of the ways in which the 'social' defines the 'interior' realm of experience and an understanding of 'how the dominant discourses within a given local site determine what can and cannot be felt and / or expressed'" (p. 231). For Boler, unearthing the obstacles and confronting the challenges to a critical articulation of emotions is essential to overcoming the "psychic imperialism" (p. 228) that implicates us all in exploitation and injustices. As Boler points out, no imperialist project has been possible without a deliberate emotional "engineering" of the polity, one that shapes emotional engagement and apprehension for exploitation and political profit.

Both Judith Butler and Megan Boler make these points about sensate democracy and critical emotional literacy while writing about and in the context of the imperialist wars of the

United States in Iraq during the 1990s and the wars in Iraq and Afghanistan following the attacks on the United States of September 11, 2001. Yet, the regulation of grief and the management of loss for nationalistic and political purposes, to which both scholars differently point, is a controlling strategy not confined to wars or to specific nations. The denial of loss and the refusal of mourning are strategies of control that encourage melancholy and a culture of blame and revenge or shame and helplessness—blunt or muted emotional conditions that are antithetical to healing. They also inhibit the development of a sensate democracy in which new visions might be created and new ways of being together in difference forged. This loss—of an ability to dream together and to be better together—may well be the greatest loss of all.

Notes

Introduction

1. A shortened version of "all at sea," a phrase used in the 1700s, prior to the development of advanced navigational instruments, to describe the precarious and uncertain state of a boat beyond sight of land and in danger of becoming lost. In contemporary usage, the shortened version, "at sea," metaphorically describes a loss of direction, of becoming unmoored and disoriented, and without a secure position.
2. When referring to the province and its people, I use the full and official name, Newfoundland and Labrador, adopted in 2001. The only exceptions are documented references to the place that precede or ignore the official formation. Where references to Newfoundland or to Newfoundlanders occur, they refer specifically to the island portion of the province, only, and to those people who have inhabited or do inhabit it. Likewise, reference to Labrador and Labradorians refer specifically to the mainland portion of the province and to those people who have inhabited or do inhabit it.
3. It is estimated that, among the current population of the province, only one person in one hundred is categorized as a member of a "visible minority." The province attracts less than 0.2 percent of all immigrants to Canada, approximately 500 persons per annum (*The Newcomer*, 2006).
4. The word *migration* is used throughout this document to denote the general movements of peoples across various kinds of constructed (and restrictive) political borders and lines—province, state, region, nation. In a context of global *transnationalism*, defined as "the flow of people, ideas, goods, and capital across national territories in a way that undermines nationality and nationalism as discrete categories of identification, economic organization, and political constitution" (Braziel & Mannur, 2003, p. 8), it is more widely useful than the more limiting *emigration* and *immigration*, used in this document only to denote the leaving or entering of a specific nation state. The phrase *outward migration* is used to describe the movement of people, in this case, from the province of

Newfoundland and Labrador, usually to another part of the country—Canada—of which it is a part.
5. 2001, pp. 22–23.
6. 2007, p. A-15.
7. The words *implication, complicity* and *collusion* are etymologically close. All three suggest a bringing or folding together, an interweaving and entangling. In their more contemporary usages, all three can also suggest secrecy and suspicion. But, here, I am more interested in implication as a series of relationships, a relation of one thing to another that brings about a form of involvement that requires a suspicious scrutiny and self-questioning. *Implication*, in this sense, captures a deep and profound connectedness.
8. The story of Lanier Phillips, an African American sailor thrown ashore in St. Lawrence when the USS *Truxton* was shipwrecked off the coast in 1942, is a story that intersects with the desires of this claim. Phillips had grown up in the segregated U.S. South and, when the ship was being abandoned, he feared that, as a black man, he would surely die if he were to go ashore. Instead, he was cared for and accepted by the people of St. Lawrence and, to this day, he credits their treatment of him at that time to changing his life. Phillips went on to challenge segregation and the racism out of which it was formed. Phillips has visited Newfoundland and Labrador often, to thank the people of St. Lawrence and, most recently, in the Spring of 2008, to receive from Memorial University of Newfoundland an honorary Doctor of Laws. His story is undoubtedly compelling and inspirational. But, it is still for all taken up in a cultural context in which tensions of cultural difference can be too often denied rather than confronted. Part of the popularity of the story may also be related to its redemptive character and the manner in which it feeds a denial of difference while it also reiterates a well-known and near mythic notion of provincial character.
9. See the discussion by Paul Gilroy (2005) of national melancholia as it relates to the United Kingdom and, in relation to the United States, the discussion by Judith Butler (2004).

Chapter 1

1. 1999, p. 9.
2. Excellent examples include Dionne Brand, *The Map to the Door of No Return*; Richard Rodriguez, *Hunger of Memory*; Eva Hoffman, *Lost in Translation*; Lisa Appignanesi, *Losing the Dead*; Jamacia Kincaid, *My Brother*; and, Ruth Behar, *The Vulnerable Observer*.
3. Here, I am using *transmigration* in both senses of the word: the physical movement of bodies from one polity to another and the spiritual movement of souls from one corporeal entity to another.

Chapter 2

1. This particular usage of *queer* points both to its colloquial meaning of "unusual" or "odd" and to its current usage as an umbrella reference to persons who are gay, lesbian, bisexual, or transgendered.
2. O'Toole has written for two Canadian publications, *Maclean's* magazine and *The Globe and Mail* newspaper, as well as *The New York Times*.
3. This metaphor, aside from its cultural affinity, attempts to capture a notion of identity as "reel," that is, as unfolding in a dance between both clarity and unpredictability, inheritance and invention, an energetic, forward-looking dance that embraces hope, resists melancholy, and encourages a joy created through encounter-in-living.
4. 1996, p. 72.
5. When announcing a childbirth incentive as part of his election platform strategy to stem so called "out-migration" from the province, Williams explained, "We can't be a dying race." The comment was circulated widely and debated hotly in various media.
6. This reference is affiliated with a common reference to the Irish as a race, too. Breda Grey (2004) discusses anti-Irish racism, noting its roots in colonization and the position of the Irish migrant worker in relation to the British nation.
7. In a June 3, 2008, public forum at which it released its policy statement, the Government of Newfoundland and Labrador announced a series of nine town hall discussions to be held at various locations around the province. The purpose of the discussions is to educate and to raise awareness about cultures so that stereotypes are challenged and countered. The province hopes to attract between 1200 and 1500 new immigrants annually within five years, with a hoped-for retention rate of 70 percent (*The Newcomer*, 2007).
8. In its first term of office, the Williams Government undertook an extensive "rebranding" project to replace the colonial coat-of-arms with a more contemporary digitally stylized depiction of the provincial flower, the pitcher plant, re-presenting it as a symbol of pride, creativity, resilience and rootedness—"a symbol of all that we are" (*The Pitcher Plant*. Brand Information Pamphlet, Government of Newfoundland and Labrador, 2006). Interestingly, O'Toole describes the provincial flower as "shame-coloured" plants, "their heads hanging abjectly toward the marshy ground" (O'Toole, 1994, p. 51).
9. The "reasonable accommodation" debate on immigration and Quebec culture, formalized by a provincial commission led by sociologist Gerard Bouchard and philosopher Charles Taylor in 2007, is the most recent example of identity as a "feasible goal." The report, released in May 2008, makes thirty-seven recommendations, the overall aims of which are to preserve secularism, foster cultural harmony, and build interculturalism (Seguin, 2008).

10. A phrase from the poem *Two Loves*, by Lord Alfred Douglas, published in 1896.
11. A quote from George Eliot's *Middlemarch*, used by O'Toole in his memoir.
12. The medieval city of Santiago de Compostela, also known as the Way of St. James, in Galacia, the northern tip of Spain, was the ancient "end of the world," inhabited, in those times, by Celts, the ancestral culture of O'Toole's family and of most of the population of Ireland.
13. The extent and nature of the connection between Ireland and Newfoundland and Labrador is not uncontested. When Russell Wangersky (2007) penned a newspaper column that challenged the depth and resemblance of the two places, he was soundly rebuked by readers. One reader, Liam McErlean, formerly of Ireland and now living in the province, captured the spirit of most of the letters to the editor: "Do [the people of Newfoundland and Labrador and Ireland] have a connection that transcends geography, weather, and culture? Yes, my cousins, there is: a history of being treated like second-class citizens, of strong (and not always beneficial) influence of the church, of hard work and hard play, of dreams dashed and resources squandered....of pride in family, accomplishments and belief in a brighter future as well as a determination to stand up and be counted (2007, p. A11).
14. Due to widespread poverty, disease and large-scale emigration, in the hundred years following "Black '47," the peak of the Irish Famine, the population of Ireland decreased by over half, from more than 8 million in the mid-1840s to 4 million by mid-twentieth century.

Chapter 3

1. New Literacy Studies is based on the assumption that representational practices of meaning—reading, writing, viewing, speaking, and so on—are political and are best understood within the social, cultural, and economic contexts in which they occur. See J. P. Gee, *The New Literacy Studies and the "Social Turn,"* and Brian Street, *What's "New" in New Literacy Studies?* for a fuller articulation of New Literacy Studies.
2. The verse novel is a short narrative, told in verse form, that employs short sections offering alternating perspectives.
3. See http://www.annandseamus.com for extensive information about this operatic adaption of the verse novel.
4. *Sponsor* is a complex, multilayered concept used in relation to those who carry and convey the ideological goods to which others seek access. For a detailed discussion of the notion of literacy sponsor, see D. Brandt (2001), Sponsors of Literacy.
5. Seamus, Irish version of James, from the Hebrew Ya`kob or Jacob, meaning "supplanter" or "usurper," one who replaces.
6. Ann, from Hebrew, meaning "favour" or "grace," "prayer."

7. Kevin Major's first novel, published in 1978, is entitled *Holdfast*.
8. Statistics Canada estimates the population of Newfoundland and Labrador for 2006 was 509,677, a decline of 4,285 from the previous year. It is the fourteenth consecutive year of a population decline that began with the cod moratorium of 1992. In 2006, the death rate surpassed the birth rate.

CHAPTER 4

1. This chapter is a revised version of "Where Biography Meets Ecology: Melancholy and Transformative Education," in M. Gardner & U. Kelly, 2008.
2. I use this term, *unbelonging*, to capture the sense of becoming undone through the ongoing process of critique of identity attachments.
3. The Newfoundland and Labrador cod fishery was closed in 2003, a little more than a decade following the initial moratorium and the same year the Atlantic Cod (Newfoundland and Labrador population) was listed as an "endangered species" by the Committee on the Status of Endangered Wildlife in Canada (COSEWIC). The stocks had declined by 97 percent in the three decades prior to these announcements.
4. Claire Mowat published her own account of the Mowats' time in Newfoundland in a memoir entitled *The Outport People* (Toronto: McClelland and Steward, 1983).
5. *A Whale for the Killing* continues to impact contemporary audiences. It has been reprinted, revised, and reissued. In 1981, it was adapted as a made-for-television movie, directed by Richard Heffron. It is referenced regularly in internet discussion "blogs" of environmental concerns.
6. While there is some debate about the extent of this devastation, no one is arguing that ocean life has not been seriously compromised by global, multinational fishing practices. A recent international study led by Boris Worm of Dalhousie University in Halifax, Nova Scotia, Canada, concluded that, if present practices continue, the Earth's oceans will be completely devastated by the mid-twenty-first century (Crosbie, 2006).
7. For example, in his book *Terminal Dialogue*, political scientist Peter McKenna documents the introduction of video-lottery terminal (VLTs) gambling machines in Newfoundland and Labrador as coincident with the cod fishery moratorium. It was a time, he notes, "when the province was looking for money and people were looking for something to do" (in Porter, 2008, p. 2.). McKenna records the devastation—addictions, family breakups, suicides—that accompany gambling.
8. The final caption of the poster read: "One of the world's great resources could be wiped out if Canada does not take urgent action to protect the fish stocks of the Grand Banks from foreign overfishing. The extinction of the stocks is an environmental and economic catastrophe which must

be prevented." (Department of Fisheries, Government of Newfoundland and Labrador, 1992).
9. The 2006 Canadian census documented an 11.8 percent decrease in the population of the province since the cod moratorium of 1992, an estimated 68,000 people of a population of over 500,000.
10. The tension has been studied extensively and, in 2003, was the focus of a Royal Commission that documented in its report, *Our Place in Canada*, the many dimensions of this dissatisfaction as well as avenues for potential change.
11. Cultural representations of these identity politics vary from a proliferation of an unofficial flag, the Pink, White, and Green, historically identified with independence, to bumper stickers and T-shirts with slogans such as "Free Newfoundland," "Newfoundland Liberation Army," and "We Club Seals."
12. In the Winter of 2007, seal populations expert, Dr. Garry Stenson, a biologist at Memorial University of Newfoundland, warned that seal harvesting at present rates, combined with declining mortality and low reproduction in the population, will create a conservation concern for the seal herd in the very near future (Gillingham, 2007).
13. David Suzuki has referred to Newfoundland and Labrador as the canary in the coal mine." He noted that, with the collapse of its northern cod fishery, Newfoundland lost not only jobs but a way of life: "If any group or population should stand as a warning that we've got to take the environment much more seriously, I think it's Newfoundland [and Labrador]" (Kielley, 2007, p. 3).

CHAPTER 5

1. This chapter is a revised and extended version of a chapter in *Teaching, Loving, and Learning*, eds. Daniel F. Liston and James Garrison (New York: Routledge, 2004).

REFERENCES

Albrecht, G. (2005). "Solastalgia": A new concept in health and identity. *PAN, 3*, 41–55.
Albrecht, G. (2004). Environmental change and human health in upper hunter communities of New South Wales, Australia. *EcoHealth, 1*, 47–58.
Ang, I. (2000). Identity blues. In P. Gilroy, L. Grossberg, & A. McRobbie (Eds.), *Without guarantees: In honour of stuart hall* (pp. 1–13). London: Verso.
Ang, I. (1998). Migrations of Chineseness: Ethnicity in the postmodern world. *Mots Pluriels, 7*. Retrieved April 2, 2002, from http://www.arts.uwa.edu.au/MotsPluriels/MP798ia.html
Antle, R. (2007, March 14). 505,469: Population declines again in 2006. *The Telegram*, p. A1.
Appadurai, A. (2003). Disjuncture and difference in the global cultural economy. In J. E. Braziel & A. Mannur (Eds.), *Theorizing Diaspora* (pp.25–48). Oxford: Blackwell.
Appignanesi, L. (1999). *Losing the dead*. London: Chatto & Windus.
Barglow, R. (1999, January/February). The angel of history: Walter Benjamin's vision of hope and despair. *Tikkun*. Retrieved March 4, 2004, from http://www.tikkun.org/magazince/index.cfm/action/tikkun/issue/tik9901/article/990115.html
Behar, R. (1996). *The vulnerable observer: Anthropology that breaks your heart*. Boston: Beacon Press.
Berger, J. (1972). *Ways of seeing*. London: Penguin.
Berger, J. (1992). *Keeping a rendezvous*. New York: Vintage.
Blackmore, G. (2003). Sense of place: Loss and the Newfoundland and Labrador spirit. In Royal Commission on Renewing and Strengthening Our Place in Canada: Research Volume 1 (pp. 309–41). St. John's, NL: Queen's Printer.
Boldt, G. (2006). Resistance, loss, and love in learning to read: A psychoanalytic inquiry. *Research in the Teaching of English, 40*(3), 272–309.
Boler, M. (1997). License to feel: Teaching in the context of war(s). In A. Cvetkovich & D. Kellner (Eds.), *Activating the Global and the Local: Globalization and Cultural Studies* (pp. 226–43). Boulder, CO: Westview Press.

Boler, M. (1998). *Feeling power: Emotions and education.* New York: Routledge.
Brand, D. (2003). *A map to the door of no return.* Toronto: Vintage.
Brandt, D. (2001). Sponsors of literacy. In E. Cushman, E. R. Kintgen, B. M. Kroll, & M. Rose (Eds.), *Literacy: A critical sourcebook* (pp. 555–71). Boston: Bedford/St. Martin's.
Braziel, J. E., & Mannur, A. (2003). Nation, migration, globalization: Points of contention in diaspora studies. In J. E. Braziel & A. Mannur (Eds), *Theorizing diaspora* (pp. 1–22). Oxford: Blackwell.
Britzman, D. (2000). If the story cannot end: Deferred action, ambivalence, and difficult knowledge. In R. I. Simon, S. Rosenberg, & C. Eppert (Eds.), *Between hope and despair: Pedagogy and the remembrance of historical trauma* (pp. 27–58). Lanham, MD: Rowman & Littlefield.
Britzman, D. P. (1998). *Lost objects, contested subjects.* Albany: SUNY Press.
Brown, W., Colegate, C., Dalton, J., & Thill, C. (2006, January) Learning to love again: An interview with Wendy Brown [Electronic version]. *Contretemps,* 25–42.
Butler, J. (2008). An account of oneself. In B. Davies (Ed.), *Judith Butler in conversation: Analyzing the texts and talk of everyday life* (pp. 19–38). New York: Routledge.
Butler, J. (2008). Conversation with Judith Butler 1. In B. Davies (Ed.), *Judith Butler in conversation: Analyzing the texts and talk of everyday life* (pp. 1–17). New York: Routledge.
Butler, J. (2004). *Precarious life: The powers of mourning and violence.* London: Verso Books.
Butler, J. (2003). Afterword: After loss, what then? In D. L. Eng & D. Kazanjian (Eds.), *Loss: The politics of mourning* (pp. 467–73). Berkeley: University of California Press.
Butler, J. (1997). *The psychic life of power: Theories in subjection.* Stanford: Stanford University Press.
Cairns, J. (2003). Reparations for environmental degradation and species extinction: A moral and ethical imperative for human society. *Ethics in Science and Environmental Politics,* 25–32. Retrieved June 20, 2007, from http://www.intres.com/articles/esep/2003/E31.pdf
Casmier-Paz, L. (2000, Summer). "Betcha can't read my mind": The orientations of literacy discourse. *Quarterly Journal of Ideology.* Retrieved June 11, 2003, from http://www.lsus.edu/la/journals/ideology/contents/casmierarticle.htm
Cheng, A. A. (2000). *The melancholy of race: Psychoanalysis, assimilation, and hidden grief.* Oxford: Oxford University Press.
Christian, M. (2001). The psychological splitting of America. *Psych Discourse,* *32*(10), 12–13.
Cohen, L. (1992). The future [Recorded by L. Cohen]. On *The future* [CD]. Don Mills, ON: Sony Music Canada.
Corbett, M. (2007). *Learning to leave.* Halifax, NS: Fernwood Books.
Crosbie, J. (2006, November 10). End in sight. *The Independent,* p. 2.

Crozier, L. (1999). *What the living won't let go*. Toronto: McClelland & Stewart.
Crummey, M. (2001). A time and place apart. *Maclean's, 114*(33), 22–23.
Crummey, M., & Furey, L. (2002). Leo Furey interviews Michael Crummey [Electronic version]. *The Antigonish Review*, 131.
Davies, B. (Ed.) (2008). *Judith Butler in conversation: Analyzing the texts and talk of everyday life*. New York: Routledge.
Davis, B., Sumara, D., & Luce-Kapler, R. (2000). *Teaching and learning in a complex world*. Mahwah, NJ: Lawrence Erlbaum.
De Certeau, M. (1984). *The practice of everyday life*. Berkeley: University of California Press.
Dei, G. J. (2002). Spiritual knowing and transformative learning. New approaches to lifelong learning (NALL Working Paper #59). Retrieved on January 14, 2008, from http://www.oise.utoronto.ca/depts/sese/csew/nall/res/59GeorgeDei.pdf
Devereaux, D. (2006). "Is you a Newfoundlander?" Am I a Newfoundlander? A paper presented at the *Despite This Loss* book workshop, Memorial University of Newfoundland, July.
Didion, J. (2005). *The year of magical thinking*. New York: Knopf.
Ellwood, W. (1996, July). The NI interview with Elizabeth Penashue [Electronic version]. *New Internationalist*, 281.
Eng, D. L., & Han, S. (2003). A dialogue on racial melancholia. In D. L. Eng & D. Kazanjian (Eds.), *Loss: The politics of mourning* (pp. 343–71). Berkeley: University of California Press.
Eng, D. L., & Kazanjian, D. (2003). Introduction: Mourning remains. In D. L. Eng & D. Kazanjian (Eds.), *Loss: The politics of mourning* (pp. 1–25). Berkeley: University of California Press.
Frampton, P. (2008, April 27). Too much rant, not enough rationale. *The Telegram*, p. A3.
Freire, P. (1970). *The pedagogy of the oppressed*. New York: Continuum.
Freud. S. (1989). Mourning and melancholia. In P. Gay (Ed.), *The Freud reader* (pp. 584–89). New York: W. W. Norton. (Original work published 1917)
Gallop, J. (Ed.) (1998). *Pedagogy: The question of impersonation*. Bloomington: Indiana University Press.
Gardner, M., & Kelly, U. (Eds.) (2008). *Narrating transformative learning in education*. New York: Palgrave Macmillan.
Gee, J. P. (2001). Literacy, discourse, and linguistics: Introduction *and* What is literacy? In E. Cushman, E. R. Kintgen, B. M. Kroll, & M. Rose (Eds.), *Literacy: A Critical Sourcebook*. (pp. 525–44). Boston: Bedford/St. Martin's.
Gee, J. P. (n.d.). New Literacy Studies and the "social turn." Retrieved May 29, 2008, from http://eric.ed.gov/ERICWebPortal/custom/portlets/ecordDetails/detailmini.jsp?_nfpb=true&_&ERICExtSearch_SearchValue_0=ED442118&ERICExtSearch_SearchType_0=no&accno=ED442118.

Gilbert, J. (1998). Reading colorblindness: Negation as an engagement with social difference. *Journal of Curriculum Theorizing*, *14*(2), 29–34.

Gillingham, R. (2007, January 22). Warning issued. *The Telegram*, p. A1.

Gilroy, P. (2005). *Postcolonial melancholia*. New York: Columbia University Press.

Gopinath, G. (2003). Nostalgia, desire, diaspora: South Asian sexualities in motion. In J. E. Braziel & A. Mannur (Eds.), *Theorizing Diaspora* (pp. 261–79). Oxford: Blackwell.

Grey, B. (2004). *Women and the Irish diaspora*. London: Routledge.

Gruenewald, D. A. (2003). The best of both worlds: A critical pedagogy of place. *Educational Researcher*, *32*(4), 3–12.

Grumet, M. (1998). *Bitter milk: Women and teaching*. Amherst: University of Massachusetts Press.

Guy, R. (2007, March 30). Newfoundland: Half heaven and half hell. *The Independent*, p. 5.

Hall, S. (2003). Cultural identity and diaspora. In J. E. Braziel & A. Mannur (Eds.), *Theorizing diaspora* (pp. 233–46). Oxford: Blackwell.

Hall, S., & Jacques, M. (Eds.) (1989). *New times*. London: Lawrence & Wishart.

Haslam, R. (1999). "A race bashed in the face": Imagining Ireland as a damaged child [Electronic version]. *Jouvert: A Journal of Postcolonial Studies*, *41*(1), 1–42.

Hoffman, E. (2000). Wanderers by choice. *UTNE Reader*. Retrieved October 4, 2007, from http://www.utne.com/2000-11-01/WanderersbyChoice.aspx

Hoffman, E. (1989). *Lost in translation*. New York: Penguin Books.

Holly, M. (1998). Patterns in the shadows. *Invisible Culture: An Electronic Journal of Visual Studies*, 1. Retrieved September 19, 2007, from http://www.rochester.edu/in_visible_culture/issue1/holly/holly.html

Humphreys, H. (2002). *The lost garden*. Toronto: Harper Flamingo Canada.

Huston, N. (2002). *Losing north*. Toronto: McArthur and Company.

Hutcheon, L. (1998). Irony, nostalgia, and the postmodern. University of Toronto English Library (UTEL). Retrieved July 22, 2006, from http://www.library.utoronto.ca/utel/criticism/hutchinp.html

Johnson, R. (1999). Exemplary differences: Mourning (and not mourning) a princess. In A. Kear & D. L. Steinberg (Eds.), *Mourning Diana: Nation, Culture and the Performance of Grief* (pp. 15–39). London: Routledge.

Johnson, R. (1997). Grievous recognitions 2: The grieving process and sexual boundaries. In D. L. Steinberg & R. Johnson (Eds.), *Border Patrols: Policing the Boundaries of Heterosexuality* (pp. 232–52). London: Cassell.

Kafka, F. (1904, January 27). Letter to Oskar Pollak. In email correspondence with Mauro Nervi of The Kafka Project. Web site: http://www.kafka.org.

Kear, A., & Steinberg, D. L. (1999). Ghost writing. In A. Kear & D. L. Steinberg (Eds.), *Mourning Diana: Nation, Culture and the Performance of Grief* (pp. 1–14). London: Routledge.

References

Kelly, U. A. (2003). The place of reparation: Love, loss, ambivalence and teaching. In D. Liston & J. Garrison (Eds.), *Teaching, Learning and Loving* (pp. 153–67). New York: RoutledgeFalmer.

Kelly, U. A. (1997). *Schooling desire: Literacy, cultural politics, and pedagogy.* New York: Routledge.

Kelly, U. A. (1993). *Marketing place: Cultural politics, regionalism and reading.* Halifax: Fernwood Books.

Kessler, R. (2004). Grief as a gateway to love in teaching. In D. Liston & J. Garrison (Eds.), *Teaching, learning and loving* (pp. 137–52). New York: RoutledgeFalmer.

Kessler, R. (2000). PassageWorks Institute: Connection, compassion and character in learning. Retrieved on June 3, 2008, from www.passageworks.org

Kielley, K. (2007, February 1–7). Suzuki embraces Newfoundlanders. *The Express*, p. 3.

Kincaid, J. (1997). *My brother.* New York: Farrar, Straus and Giroux.

Kincheloe, J. L. (2005). *Critical constructivism.* New York: Peter Lang.

Klein, M. (1975). *Love, guilt and reparation and other works 1921–1945.* London: Hogarth Press.

Kristeva, J. (1980). *Desire in language* (L. S. Roudiez, Ed. and Trans.). New York: Columbia.

Lear, J. (2006). *Radical hope: Ethics in the face of cultural devastation.* Cambridge: Harvard University Press.

Lloyd, D. (2003). The memory of hunger. In D. L. Eng & D. Kazanjian (Eds.), *Loss: The politics of mourning* (pp. 205–28). Berkeley: University of California Press.

Luke, A. (2003). Literacy education for a new ethics of global community. *Language Arts, 81*(1).

Luke, A., & Carrington, V. (2004). Globalization, literacy, curriculum practice. In T. Grainger (Ed.), *The RoutledgeFalmer reader in language and literacy* (pp. 52–65). London: RoutledgeFalmer.

Mac Einri, P. (2001). States of becoming: Is there a "here" here and a "there" there? Retrieved April 2, 2002, from http://migration.ucc.ie/statesofbecoming.htm

Major, K. (2003). *Ann and Seamus* (David Blackwood, Ill.). Toronto: Groundwood.

Major, K. (2001). *As near to heaven by sea: A history of Newfoundland and Labrador.* Toronto: Penguin/Viking.

Malone, G. (2007). On happiness. *Newfoundland Quartlerly, 99*(3), 45–47.

Marris, P. (1986). *Loss and change.* New York: Routledge.

Martusewich, R. (1997, Fall). Leaving home: Curriculum as translation. *Journal of Curriculum Theorizing, 13*(3), 13–17.

Massey, D. (2000). Travelling thoughts. In P. Gilroy, L. Grossberg, & A. McRobbie (Eds.), *Without guarantees: In honour of Stuart Hall* (pp. 225–32). London: Verso.

Massey, D. (1994). *Space, place and gender.* Cambridge: Polity Press.

McErlean, L. (2007, December 1). A sea of difference [Letter to the editor]. *The Telegram*, p. A11.
Michaels, A. (1997). *The weight of oranges / Miner's pond*. Toronto: McClelland & Stewart.
Mitchell, J. (Ed.) (1986). *The selected Melanie Klein*. New York: Free Press.
Moore, S. (2005). Mourning, melancholia, and death drive pedagogy: Atwood, Klein, Woolf. *Journal of the Canadian Association for Curriculum Studies, 3*(2), 87–102.
Morgan, B. (2007). *Cloud of bone*. Toronto: Knopf Canada.
Mowat, F. (2008, April 16). Two against the tide? *The Globe and Mail*, p. A13.
Mowat, F. (1972). *A whale for the killing*. Toronto: McClelland and Stewart.
Munro, A. (2006). *The view from Castle Rock*. Toronto: McClelland & Stewart.
Murphy, R. (2007, September 1). So gloomy the future, it tears my rock. *The Globe and Mail*, p. A15.
Newfoundland and Labrador launches provincial immigration strategy (2008). *The Newcomer: A Quarterly Newsletter on Immigration in Newfoundland and Labrador, 1*(4).
Oliver, M. (2004). *Long life: Essays and other writings*. Cambridge, MA: Da Capo Books.
Ong, W. (2001). Writing is a technology that restructures thought. In E. Cushman, E. R. Kintgen, B. M. Kroll, & M. Rose (Eds.), *Literacy: A critical sourcebook* (pp. 19–31). Boston: Bedford/St. Martin's.
Ortega, E. (2005, October). Review of *Ann and Seamus*. *School Library Journal, 168*.
Ostroff, M. (2006). *To think like a composer—The making of Ann and Seamus: A Chamber Opera*. St. John's, NL: Rink Rat Productions and Cine Metu.
O'Sullivan, E. (2008). Notes towards a transformative education. In M. Gardner and U. Kelly (Eds.), *Narrating transformative learning in education* (pp. ix–xvii). New York: Palgrave Macmillan.
O'Sullivan, E. (2002). The project and vision of transformative education: Integral transformative learning. In E. O'Sullivan, A. Morrell, & M. O'Conner (Eds.), *Expanding the boundaries of transformative learning* (pp. 1–12). New York: Palgrave.
O'Toole, L. (1994). *Heart's Longing: Newfoundland, New York and the distance home*. Toronto: Douglas and McIntyre.
Overton, J. (1996). *Making a world of difference*. St. John's, NL: ISER Books.
Pandurang, M. (2001). Cross-cultural texts and diasporic identities. Retrieved April 4, 2002, from http://social.chass.ncsu.edu/jouvert/v6i1-2/bromle.htm
Panting, S. (2007, March 30). Poor pitiful us. *The Independent*, p. 19.
Phelan, A. M. (2003). Melancholia. *Educational Insights. 8*(2).
Pinar, W. F. (1991). Curriculum as social psychoanalysis: On the significance of place. In J. L. Kincheloe & W. F. Pinar (Eds.), *Curriculum as social*

psychoanalysis: The significance of place (pp. 165–86). Albany, NY: SUNY Press.
Pitt, A. (2006). Mother love's education. In G. M. Boldt & P. M. Salvio (Eds.), *Love's return: Psychoanalytic essays on childhood, teaching, and learning* (pp. 87–105). New York: Routledge.
Pitt, A., Robertson, J. P., & Todd, S. (1998). Psychoanalytic encounters: Putting pedagogy on the couch. *Journal of Curriculum Theorizing, 14*(2).
Porter, S. (2008, April 4). "Culture of shame." *The Independent*, pp. 1–2.
Porter, S. (2005, January 23). Rex Murphy says Newfoundlanders are fighting for cultural survival [online version]. *The Independent*.
Probyn, E. (1996). *Outside belongings*. New York: Routledge.
Probyn, E. (1995). Queer belongings: The politics of departure. In E. Grosz & E. Probyn (Edss), *Sexy bodies: The strange carnalities of feminism* (pp. 1–18). New York: Routledge.
Progressive Conservative Party Kicks of Election 2007 (2007, September 17). Retrieved October 9, 2007, from http://www.pcparty.nf.net/200709171.htm
Proulx, E. A. (1993). *The shipping news*. New York: Touchstone.
Reid, T. (1992) (Dir.). *Memory waltz* [music video]. Halifax: Ground Swell Records.
Rodriguez, R. (1983). *Hunger of memory: The education of Richard Rodriguez*. New York: Bantam Books.
Ross, V. (2006, October 28). Lunch at Alice's Restaurant. *The Globe and Mail*, p. R6.
Royal Commission on Renewing and Strengthening Our Place in Canada. (2003). *Our place in Canada*. St. John's, NL: Queen's Printer.
Ryan, D. W. S., & Rossiter, T. P. (Eds.) (1984). *The Newfoundland character: An anthology of Newfoundland and Labrador writings*. St. John's, NL: Jesperson Press.
Salvio, P. (2006). On the vicissitudes of love and hate: Anne Sexton's pedagogy of loss and reparation. In G. M. Boldt & P. M. Salvio (Eds.), *Love's return: Psychoanalytic essays on childhood, teaching, and learning* (pp. 65–87). New York: Routledge.
Salvio, P. (1998a). On using the literacy portfolio to prepare teachers for "willful world travelling." In W. Pinar (Ed.), *Curriculum: Toward New Identities* (pp. 41–74). New York: Garland Press.
Salvio, P. (1998b). The teacher/scholar as melancholic: Excavating scholarly and pedagogic (s)crypts in *Fugitive pieces. Journal of Curriculum Theorizing, 14*(2), 15–23.
Samson, C., Wilson, J., & Mazower, J. (1999). *Canada's Tibet: The killing of the Innu*. London, UK: Survival International.
Sanchez-Pardo, E. (2003). *Cultures of the death drive*. Durham, NC: Duke University Press.
Schaefer, J. (2006, December 30). Vibrant economy is taking a toll on wildlife. *The Telegram*, p. A11.

References

Schumacher, J. F (2006, July) The happiness conspiracy, *New Internationalist*, 393.

Scribner, S. (1986). Literacy in three metaphors. *Journal of American Education*, 93, 6–21.

Sedgwick, E. K. (2003). *Touching feeling: Affect, pedagogy, performativity.* Durham: Duke University Press.

Sedgwick, E. K. (1999). *A dialogue on love.* Boston: Beacon Press.

Seguin, R. (2008, May 23). Quebec's day of reckoning. *The Globe and Mail*, pp. A1, A8.

Sheldrake, P. (2001). *Spaces for the sacred: Place, memory, and identity.* Baltimore: Johns Hopkins University Press.

Shreve, A. (1997). *The weight of water.* New York: Little Brown.

Simon, R. I. (1993). *Teaching against the grain: Texts for a pedagogy of possibility.* Toronto: OISE Press.

Simon, R. I., Rosenberg, S., & Eppert, C. (2000). Introduction. In R. I. Simon, S. Rosenberg, & C. Eppert (Eds.), *Between hope and despair: The pedagogical encounter of historical remembrance* (pp. 1–8). Lanham, MD: Rowman & Littlefield.

Spivak, G. C. (2000). Thinking Cultural Questions in "Pure" Literary Terms. In P. Gilroy, L. Grossberg, & A. McRobbie (Eds.), *Without guarantees: In honour of Stuart Hall* (pp. 344–45). London: Verso.

Steedman. C. (1986). *Landscape for a good woman.* London: Virago.

Street, B. (2003). What's "new" in new literacy studies? *Current Issues in Comparative Education.* 5(2),77–91.

Street, B. (2001). The new literacy studies. In E. Cushman, E. R. Kintgen, B. M. Kroll, & M. Rose (Eds.) (2001). *Literacy: A critical sourcebook* (pp. 430–42). Boston: Bedford/St. Martin's.

Tilley, S. A. (2000). Provincially speaking: "You don't sound like a newfoundlander." In C. E. James (Ed.), *Experiencing difference* (pp. 235–45). Halifax, NS: Fernwood.

Todd, S. (1997). *Learning desire: Perspectives on pedagogy, culture, and the unsaid.* New York: Routledge.

Walcott, R. (2000). Pedagogy and trauma: The middle passage, slavery, and the problem of creolization. In R. I. Simon, S. Rosenberg, & C. Eppert (Eds.), *Between hope and despair: The pedagogical encounter of historical remembrance* (pp. 135–51). Lanham, MD: Rowman & Littlefield.

Walden, S. (2001). Ghost ports: A photo essay. *Maclean's*, 114(33), 22–28.

Wangersky, R. (2007, November 24). Ire and the Irish connection. *The Telegram*, p.A11.

Whitlock, G. (1994). White diasporas: Joan (and Ana) make history. Retrieved April 2, 2002, from http://www.arts.uwo.ca/~andrewf/anzsc12/whitlock12.htm

Williams, R. (1992). *The long revolution.* London: Hogarth.

Williams, R. (1973). *The country and the city.* London: Chatto and Windus.

Winterson, J. (1995). *Art objects: Essays on ecstasy and effrontery*. London: Jonathan Cape.

Wright, E. (1984). *Psychoanalytic criticism: Theory into practice*. London: Methuen.

Wright, R. (2004). *A short history of progress*. Toronto: House of Anansi.

INDEX

aboriginal peoples, 6, 11, 39, 43–44, 104–5, 154
AIDS, 49, 61, 63, 67
Albrecht, Glenn, 6, 103–4, 105
ambivalence, 9, 15; and melancholia, 9; and migration, 62–63; and teaching and learning, 20
Ang, Ing, 14, 57, 98, 109, 165–66
Ann and Seamus, 20, 72, 130, 157, 161; binaries in, 81–83; contemporary relevance of, 74, 76; historical background of, 72, 87; literacy and narrative structure of, 73; as reparative gesture, 96–97; synopsis of, 73; as verse novel, 72
Ann and Seamus: A Chamber Opera, 74, 158
Antle, Rob, 30
Appadurai, Arjun, 44
Appignanesi, Lisa, 139
Art Objects, 137
As Near to Heaven by Sea, 87
assimilation: cultural subjects of, 51–53
at sea, 1; meaning, 4, 169
attachment, 151; affective dimensions of place, 100–110; place and, 74–76, 78–81, 89–90, 94, 111–12, 116–17, 130–37; and separation, 117, 131, 151–52
Australia, 57, 103

autobiography, 5, 19, 48, 122–23, 164

Barglow, Ray, 87
Behar, Ruth, 24, 26, 41
belonging: narratives of, 67–68, 91; new forms of, 18, 67–68, 91, 165; sexual identity and, 58–61; and unbelonging, 93, 159, 163, 173
Beothuck, 12, 36, 157, 161
Berger, John, 5, 23, 44, 51
Bitter Milk, 111
Blackmore, Ged, 36, 37, 42
Blackwood, David, 20, 72, 76; and melancholic images, 85
Boldt, Gail, 80, 80–81, 84, 88, 122, 159
Boler, Megan, 13, 17, 138, 147, 160, 167
Brand, Dionne, 12, 93, 133–34, 135, 158
Brandt, Deborah, 79
Braziel, Jana Evans, 28, 31, 169
Britain, 29; imperialism, 77–78, 108
Britzman, Deborah, 122, 129, 138, 140, 149
Bromley, Roger, 5, 36, 159, 163, 164
Brown, Cassie, 98
Brown, Wendy, 42, 68, 109, 110, 116, 117–18
Bury me in Newfoundland, xi

Butler, Judith, 7, 8, 14, 17, 18, 26, 28, 37, 51, 57, 58, 84, 110, 118, 122–23, 127, 132, 148, 149, 152, 167, 170

Cairns, Jr., John, 115
Camino de Santiago, 63, 64
Carrington, Victoria, 162, 163
Casmier-Paz, Lynn, 81
Cheng, Anne Anlin, 6, 7, 8, 51, 83, 84, 110
Christian, Marcelle, 132
Cohen, Leonard, 112
Colegate, Christina, 109, 110
Corbett, Michael, 39, 90, 153
Crozier, Lorna, 27
Crummey, Michael, 2, 156
cultural mythology,130; of character, 54–55, 170; of home, 163; of homogeneity, 29, 43; of life on the sea, 77; and place attachment, 89; of struggle and survival, 118, 130
curriculum vitae, 32–34

Dalton, John, 109, 110
Davis, Brent, 84, 114
death and dying: and academia, 24–25, 27, 32–34, 41; prevalence of metaphors of, 11, 89–90
Death on the Ice, 98, 113
de Certeau, Michel, 24, 40Dei, George, 155
democracy: sensate, 149, 167–68
Devereaux, Danielle, xi, 89
Devlin, Bernadette, 68
Dialogue on Love, A, 144
diaspora, 28–31; African, 15, 133–34, 140–41; Biblical, 62; and identity, 28
diasporic peoples: in Newfoundland, 29, 43
Didion, Joan, 124, 148
Dyer Knight, Susan, 74, 157

ecology, 94; and culture, 94, 99–100; and economy, 98, 99–100; and place, 152–54; and reparation, 115–16; and solastagia, 103–7
education: childhood experiences of, 29; and ecological justice, 95–96, 151–54; and leaving, 78–79, 90–91; loss and place and, 16–18, 19, 21, 27, 40–45, 48, 143–44, 148–68, 159; as reparative, 96, 100, 114–18, 159; social and cultural difference and, 16, 127, 137, 139–41, 159, 160, 164–65
Ellwood, Wayne, 26, 154
emigration: usage, 169
Eng, David, 7, 8, 16, 24, 51, 52, 84, 108, 110, 159, 161
England, 16, 28
Eppert, Claudia, 12, 17

Frampton, Pam, 113
Freire, Paulo, 144
Freud, Sigmund, 8, 14, 51, 83, 107, 149
Furey, Leo, 156

Gallop, Jane, 122
Garrison, James, 121
Gee, James Paul, 71, 79
Gilbert, Jen, 138
Gilroy, Paul, 2, 8, 51, 149, 161, 162, 170
globalization, 2, 31, 94, 112–13, 160, 162–65
Globe and Mail, The, 31
Gopinath, Gayatri, 58
Greenpeace, 98, 113
Grey, Breda, 171
grief: reflective, 19, 25, 121, 149, 150–51; and identity, 124; as normative, 25–26, 150–51; and sexuality, 48; as social,

25–26; within transformative education, 95–96; work, 25, 27
Gruenewald, David, 115, 153
Grumet, Madeline, 111, 129, 160, 162
Guy, Ray, 42

Hall, Stuart, 2, 54, 56, 163
Han, Shinhee, 9, 16, 24, 51, 52, 159, 161
Harvey, Ann: historicial figure, 72–73
Haslam, Richard, 65
Hatfield, Stephen, 74, 158
Heart's Longing: Newfoundland, New York, and the Distance Home, 19, 47–51, 77, 150; characters, 50, 59; as memoir of loss and mourning, 51–68; sexual identity and, 58–61; synopsis, 49
Hoffman, Eva, 51
Holly, Michael Anne, 64
Humphries, Helen, 27, 124, 142
Huston, Nancy, 4, 39, 53, 63
Hutcheon, Linda, 101, 156

identity: and change, 14, 165–67; and diaspora, 28; and disavowal, 57; ecological postmodern perspective on, 114; and education, 114; as essence, 54–56; fortress, 98; and loss, 26, 27; as position, 54, 56–57; and provincial rebranding, 171; as reel, 53, 67–68, 165, 171; repudiation of, 58–60; and sexuality, 48, 56, 57, 58–61; vectors of, 56, 67
Identity Blues, 57
immigration: usage, 169
implication, 7, 12, 14, 110, 119, 152; in ecological crisis, 94, 99–100, 105–7; and education, 114, 129, 162; meaning and usage, 170
Innu, 6, 26, 104, 154
Ireland, 16, 28, 69, 108
Irish, 29, 64, 77
Irish Famine, The, 64, 65, 69, 172

Johnson, Richard, 25, 26, 28, 35, 37, 124, 125, 126

Kafka, Franz, 155
Kazanjian, David, 7, 8, 84, 108, 110
Kear, Adrian, 23, 30
Kelly, Ursula, 5, 31, 40, 97, 108, 162
Kessler, Rachael, 24, 28, 41, 42, 147, 148
Kincaid, Jamaica, 61, 62
Kincheloe, Joe, 25
Klein, Melanie, 14, 15, 16, 21, 37, 61, 64, 115, 123, 124, 125, 126, 127, 132, 137–38; depressive position, 37, 125–26; displacement, 132–33, 135; idealization, 139–40
Kristeva, Julia, 84, 88

Labrador, 26, 98, 154; usage, 169
Labradorians, 39, 55
Landscape for a Good Woman, 110
Lear, Jonathan, 13, 41, 111–12, 156
learning: childhood experiences of, 98, 111–12; and loss, 17; and mother-daughter relationship, 127–28
Liston, Daniel, 121
literacy, 3, 20, 71; alphabetic, 82, 84–85, 88; critical, 3, 75, 162–65; critical emotional, 147, 167–68; colonial discourse and, 81, 88–89; conviviality and, 162–65; as cultural demand, 73, 88, 157; identity kit, 79; literacy crisis and, 71,

literacy *(continued)* 81; and loss, 79–81, 90–91; New Literacy Studies and, 71, 72; primary discourse in, 79–80; psychoanalysis and, 72, 84–89; psychosocial aspects, 20, 72; sponsor, 79, 172; and shame, 88–89; as state-of-grace, 81; and written language, 81–83

Lloyd, David, 108

Long Life, 96, 100

Losing North, 4

loss: acknowledgment of, 148–49; and community response, 67; as cultural, 2, 4, 18, 26, 36–40; cultural narratives of, 67–68, 155–58; and environment, 20; and sexuality, 20, 48; and unarticulated grief, 7, 36–40

Lost Garden, The, 27, 124, 142

Luce-Kapler, Rebecca, 84, 114

Lugones, Maria, 164

Luke, Allan, 162, 163, 164

Mac Einri, Piaras, 38, 43, 166

Major, Kevin, 20, 72, 73, 79, 80, 85, 86, 87, 88

Malone, Greg, 106, 120

Manhattan, 49, 67

Mannur, Anita, 28, 31, 169

Marketing Place, 3, 22

Marris, Peter, 37

Martusewicz, Rebecca, 117

Massey, Doreen, 135, 136, 163, 164

Mazover, Jonathan, 154

melancholia, 6, 20, 83; and ambivalence, 9, 62, 83–84; and assimilation, 51–53, 84; as critical tool, 9, 17; and identity, 52, 83–84; meaning, 6; and melancholy, 6; as misapprehension, 10, 149, 167–68; and mourning, 8, 51–52, 83–84; and nationalism, 152, 168, 170; and oppression, 9–10; and place, 10–12, 107–10; as process of negotiation, 9; and reparation, 13–16, 52; and sexuality, 60; as symptom of assimilation, 9

Memory Waltz, 150

Mercer, Kobena, 2, 160

Michaels, Anne, 124

migrant, 9, 23, 52, 66, 157–158

migrant story, 23–24, 26, 27–40, 44; conventions of, 30; social relations and, 31; in teacher education, 44–45, 164–65

migration, 2, 133–34; culture of, 24; differences among, 160–61; economic, 30; and identity, 5, 35–36, 91, 101–3; and loss, 5, 19, 38, 48, 49; and nostalgia, 101–3; and return, 35–40, 52–53, 134–37; sexuality and, 58, 62; usage, 169

Miller, Alice, 66, 67

Mitchell, Juliet, 61

Morgan, Bernice, 161

Mount Cashel Roman Catholic Orphanage, 10, 36

mourning: and migration, 24, 48; psychic processes, 14, 83; as reparation, 126

Mourning and Melancholia, 8, 83

Mowat, Claire, 96

Mowat, Farley, 96–98

Munro, Alice, 61

Murphy, Rex, 2, 13

My Brother, 61

Natuashish, 154

Newcomer, The, 169, 171

Newfoundland, 29, 32, 33, 56, 64; usage, 169

Newfoundland Character, The, 54

Newfoundlander, 33, 34, 36, 39, 40, 55, 58, 99, 137

Newfoundland Youth Symphony Choir, 157. *See also* Shallaway
Newfoundland and Labrador, 2; cod fishery, 3, 10, 11, 14, 30, 36, 38, 49, 85–86, 89, 94, 99, 105–7, 150, 173–74; Confederation with Canada, 10, 33, 98, 109, 154; cultural catastrophe and, 94, 99–100, 106–7, 116–17; cultural losses and, 10, 39, 42; cultural stereotypes, 11, 34, 102–3; current context of change, 2–4; Government of, 55; and historical disavowals, 12, 161–62; identity as essence, 54–56; identity as position, 56–57; and immigration, 2, 42–43, 166, 169, 171; and Ireland, 38–39, 64–66, 108, 172; and melancholic attachments, 11; and migrant story, 30–31; population shifts, 30, 173, 174; seal hunt, 97, 98–99, 113; usage, 169; as White settler colony, 2, 11, 31, 43
nostalgia, 5, 20, 35–40, 58, 100–103
Notes Towards a Transformative Education, 152
Nova Scotia, 105, 153

Office of Immigration and Multiculturalism, 43, 55
oil and gas industry, 3, 11, 21
Oliver, Mary, 96, 100
Ong, Walter, 84, 91
Ortega, Eve, 79
Ostroff, Michael, 157
O'Sullivan, Edmund, 24, 100, 105, 115, 116, 117, 152, 155; integrative transformative education, 95–96, 147, 153

O'Toole, Lawrence, 19, 47, 48, 49, 50, 53, 77, 126, 150, 171
Our Place in Canada, 11
outward migration, 10, 31, 38, 43, 49, 63, 86, 90, 94, 106, 134, 166, 169, 171; as experience of loss, 25
Overton, James, 5, 101

Pandurang, Mala, 163, 164
Panting, Sean, 42
pedagogy, 15; and identity, 15–16, 113–14; and implication, 113; and reparation, 140–41, 143
Penashue, Elizabeth, 104
Phelan, Anne, 148
Phillips, Lanier, 170
Pinar, William, 42, 131, 132, 139, 166
Pitt, Alice, 128, 129, 142
Porter, Stephanie, 14
postscriptum, 19, 45, 68–69, 91–92, 119–20, 145
presentism, economic, 4, 100
Probyn, Elspeth, 53, 56, 58
Proulx, E. Annie, 130–31
Psychic Life of Power, The, 58

Quebec, Province of, 56; identity of, 56; reasonable accommodation debate in, 171
queer, 48, 58, 59, 60, 62; usage, 48, 171

Rawlins Cross, 150
Rayner, Timothy, 109, 110
reparation, 2, 14, 21, 95, 109–10, 124–26, 158–59; and ambivalence, 14, 125; and melancholia, 13–16, 52; and place, 96; and self-acceptance, 16, 141; and teaching, 21, 123, 127–30, 135, 142–45; and writing, 61–64
Robertson, Judith, 128, 129

Rosenberg, Sharon, 12, 17
Ross, Valerie, 61
Rossiter, Thomas, 54, 55
Royal Commission on Renewing and Strengthening Our Place in Canada, 11, 13, 36, 174, 175
Ryan, Donald, 54, 55

Salvio, Paula, 48, 84, 122, 128, 158, 159, 162, 163–64
Samson, Colin, 154
Sanchez-Pardo, Esther, 157
Schumacher, John, 153
Scribner, Sylvia, 81
Sedgwick, Eve Kosofsky, 126, 144–45
Shaefer, James, 99–100
Shallaway, 74, 157
Sheldrake, Philip, 131
Short History of Progress, A, 116
Shreve, Anita, 27
Silan, Jonathan G., 7
Simon, Roger. I., 12, 17, 122
solastalgia, 6, 20, 103–7; and aboriginal peoples, 6, 104–5, 154; and rural Newfoundland and Labrador, 6–7, 105–7, 173
Spivak, Gayatri Chakravorty, 62, 67, 134
Steedman, Carolyn, 110
Steinberg, Deborah Lynn, 23, 30
Street, Brian, 71, 73
structures of loss, 4–7
Sumara, Dennis, 84, 114
Survival International, 154
Suzuki, David, 120

Tainter, Joseph, 116
Talamh an ÉÉisc, 64

teaching: and ambivalence, 129–30; early experiences of, 98–99, 113; and love, 142–45; and place, 131–32, 139–40; psychic dimensions of, 128–30; and reparation, 137–41, 143; self-reflexivity, 122–23, 129
Teaching, Loving and Learning, 121
Thill, Cate, 109, 110
Tilley, Susan, 34
Todd, Sharon, 122, 128, 129
Tompkins, Silvan, 88
transmigration,1, 45, 170
transnationalism, 169
Tshaukuesh, 154, 155. *See also* Penashue, Elizabeth

vulnerability, 13–14, 112, 127, 151–52, 160, 161–62

Walcott, Rinaldo, 15, 16, 134, 140–41, 159
Walden, Scott, 63
Weight of Oranges, The / Miner's Pond, 124
Weight of Water, The, 27
Whale for the Killing, A, 96–98, 99, 173
Whitlock, Gillian, 31
Williams, Premier Danny, 55
Williams, Raymond, 20, 72
Wilson, James, 154
Winterson, Jeanette, 137
Wright, Ronald, 116
writing: as reparative, 19, 48, 61–64, 66, 155–57; and separation, 84

Year of Magical Thinking, The, 124
Youth Retention and Attraction Strategy, 38

GPSR Compliance

The European Union's (EU) General Product Safety Regulation (GPSR) is a set of rules that requires consumer products to be safe and our obligations to ensure this.

If you have any concerns about our products, you can contact us on

ProductSafety@springernature.com

In case Publisher is established outside the EU, the EU authorized representative is:

Springer Nature Customer Service Center GmbH
Europaplatz 3
69115 Heidelberg, Germany

www.ingramcontent.com/pod-product-compliance
Lightning Source LLC
LaVergne TN
LVHW041955060526
838200LV00002B/21